Jai Shree Ram

Songs of Blasting
Practices in Opencast Mine Blasting

Table of Contents

Summary 7

Chapter 1: Objective of Blasting 9

Introduction 9

Optimum Blasting curve 11

MCQ's 11

Chapter 2: Explosives 13

Definition 13

History 14

ANFO 15

SME explosive 15

Raw Material used in bulk explosives (SME) 17

Ammonium Nitrate (AN) 17

Calcium Nitrate (CN) 18

Other Nitrate Salts and additives 18

Fuel Phase 18

Emulsifier and Co-emulsifier: 19

Parameters of oxidizer solution 20

Parameters of Fuel phase 21

Parameters of Emulsion matrix 21

Chemistry of Explosives   21

Various Properties of Explosive: 24

     1: Density   24

     2: Velocity of Detonation (VOD)   26

     Steady state VOD: 32

     3. Critical Diameter   32

     4: DETONATION AND BOREHOLE PRESSURE   32

     5: Water Resistance   33

     6: Sensitivity: 33

     7: Sensitiveness: 34

     8: Fumes   34

     9: Strength   35

     BMD vehicle: 36

Multiple Choice Questions   40

Chapter 3: Properties of Rock Mass   55

     Rock Strength: 55

     Rock Density   55

     Elastic Properties of rock   56

     Wave properties of rocks   57

     Presence of Discontinuity- fault planes   57

     Presence of cavity   57

     Variation in strata in a bench: 58

     Presence of pre-formed boulders in the strata   59

     Presence of water   60

     Reactive ground   60

Theory of rock breaking during blasting   61

Blast Design Parameters   63

     Bench Height   63

     Blast hole Diameter   63

     Burden   64

     Spacing   65

     Sub Grade Drilling   67

     Blasthole length   68

     Stemming Column and charge column   68

     Deck Charging   68

     Volume Calculation   69

     Powder Factor Calculation   69

     Bench Stiffness Ratio   72

     Hole Inclination: 72

Drilling   72

Accessories used in Blasting   73

Initiating system   74

     Detonating Cord/Detonating Fuse (DF)   75

     Detonators   76

     Detonating Relays: 78

     MS Connectors   78

     NONEL   79

     Electronic Detonator   81

     Electric Detonator and Safety fuse   84

Primer   85

CARTRIDGE   85

   Cast Boosters   86

   Emulsion Booster: 87

Other Accessories   87

   Blasting Cable   87

   Exploder   87

   Measuring Tape   88

   Measuring Beaker   88

   Spring Balance: 88

   Spades and Shovels for stemming   88

   Tamping Rods   88

   Blasting Knifes or pliers   88

   Warning System during clearance   88

Charging Operations   88

Delay Sequencing   95

   Scattering of delay in holes and its impact: 97

   Firing Pattern   98

SAFETY IN FIELD OPERATIONS   103

   Friction: 103

   Impact: 104

   Static Charge: 106

   Heat: 109

Charging in Hot and cracked strata   111

   Charging in cracked holes   111

   Charging at Hot holes   112

Charging at Dragline and deep hole shovel benches   115

   Cast Blasting   115

   Toe formation in Dragline/shovel Bench   117

Multi seam Blasting   120

   Introduction   120

Post Blast Analysis   122

   Profile of blasted muck: 123

   Fragmentation: 124

   Importance of Drill pattern and fragmentation and Powder factor and there relation   125

   Improper Lifting   127

   Multiple Choice Questions   128

Chapter 4: Issues in blasting and measures to control   148

   Stemming ejection or blowout.   148

   Undercharging: 150

   Handling of Boulder: 151

   Disadvantages of Secondary Blasting   151

   Over charging :   152

   Ground Vibration: 152

   Fly rocks: 161

   Back Break   165

   Crystallization of AN in explosive: 166

   Low Viscosity product   166

   Lower rate of Gassing   167

Non Initiation of Bulk Explosive due to improper primer (use of Emulsion booster):   167

Emulsion Breaking   168

Alternatives of Rock Blasting   168

Multiple Choice Questions   170

Chapter 5: Record Keeping of Blast   177

Chapter 6: Legal compliances w.r.t. PESO   182

Legal compliances w.r.t. Factory Rules   183

Legal compliances w.r.t. Pollution (Air and Water) Acts   183

Legal compliances w.r.t. DGMS   186

Water Pollution   189

Air Pollution   189

Land Pollution   190

Methods of mitigation   190

Summary

This book covers blasting in Open cast mines using Ammonium Nitrate based Site Mixed Explosives and various technicalities in handling, proper utilization and various issues related with blasting.

Book is divided into number of chapters that will cover the manufacturing process of SME explosives, various properties of explosives, Usage of explosives in mines, safety precautions taken in handling of explosives, blasting performance parameters such as fragmentation, ground vibration and other post blast analysis.

Chapter 1: Objective of Blasting

Introduction

Blasting is basic operation in mining. In this operation, the intact rock mass which cannot be excavated directly by machines is loosened by blasting so that the waste or ore are easily handled, without causing any stress to excavator. The primary aim of blasting is to give optimum fragmented rock which can be easily lifted by excavators.

Too good fragmentation would require more explosive and since explosive has its own cost, so there is need of optimization of use of explosive. The main motto of optimization should not be merely fragment size handling by excavators, rather optimization must be from drilling, to blasting to handling of muck to further processing (in case of ore body) in crusher and mills. This is called mines to mill optimization and is very important consideration in blasting. There is very popular curve of optimum cost of blasting and size of fragment must be in line with that curve.

Summing all the points for objective of blasting:

1> Optimum fragmentation: Fragmentation of muck pile is such that it is easily handled by excavators and further easily crushed in crusher and onward processing of ore can be done. If the size of fragments is big, then loading time will increase, thus decreasing the productivity of excavator. Also boulder handling can lead to unplanned breakdowns of the machinery involved in excavating or crushing. Common breakdown that occurs in boulder handling is breaking or misalignment of hoist rope in rope shovels and draglines, over consumption of teeth of bucket, leakage of hydraulic due to high pressure in hydraulic excavators. Also too fine fragments would require closed drilling pattern and high explosive charge, thus increasing cost of drilling and blasting. This optimization of fragment size in mining is called mines to mill optimization.

2> Minimization of cost of blasting

3> Minimization of vibration, fly rocks, air over pressure

4> Minimization of back break, side break due to blasting

5> There should be minimum fumes generated due to blasting

6> Safety of person is must in any mining operation and this is also aim of blasting that there should be safety to all person involved in blasting operation'

If one considers the energy requirement for mining operations from Blasting to Milling then the energy required for blasting is amongst the lowest as compared to crushing, secondary crushing, milling and refining. So if cost of blasting is meagrely increased to produce better fines size fragments, the associated cost of secondary crushing and onward process can reduce significantly.

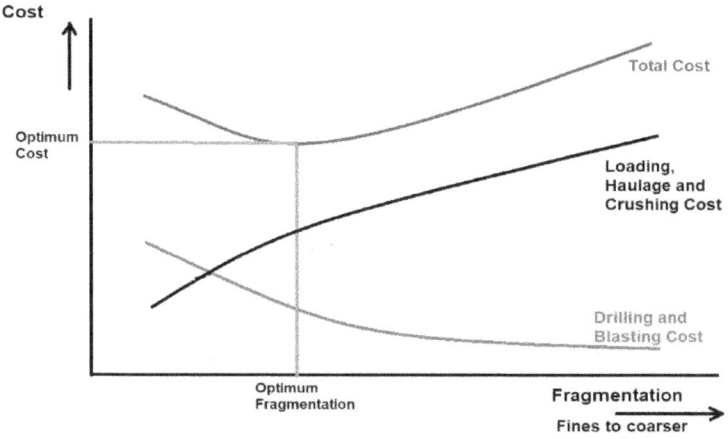

Optimum Blasting curve

As is evident from the curve, cost of loading, hauling and crushing increases when fragmentation is coarser and inversely proportional to cost of drilling and blasting. Hence when combined cost is plotted for all basic operations, the minimum cost will give the optimum cost and also give the size of fragments.

Optimum fragmentation is dependent on number of factors like bucket capacity of excavators and hauling machines, type of strata, feed size of crusher etc.

MCQ's

Q1: Which of the following is not a unit operation in mining?

A: Drilling and Blasting	C: Surveying
B: Loading and Haulage	D: Machinery Maintenance

ANS: C and D

The unit operations of mining are the basic steps used to produce mineral from the deposit, and the auxiliary operations that are used to support them. The steps contributing directly to mineral extraction are production operations, which constitute the production cycle of operations. The ancillary steps that support the production cycle are termed auxiliary operations.

Q2: Aim of Blasting is:

A: Fine fragmentation of rocks to increase load cycle and decrease mucking cost	C: Optimum fragmentation
B: Minimization of cost of drilling and blasting as explosive has high economic value	D: All of the above

Ans: C

Q3: What is suggestive from optimum blasting curve

A: Cost of Blasting decreases as fragment size decreases	C: Cost of Haulage increases as fragment size increase
B: Cost of blasting decreases as fragment size increases	D: None of the above

ANS: B

4: Optimum fragmentation depends on

A: Size of crusher feed	C: Carrying capacity of Dumper

B: Bucket capacity of excavator	D: All of the above

Ans: D

Chapter 2: Explosives

Definition

Explosive is a chemical which if subjected to required conditions burn rapidly and produces huge energy in form of shock, gas pressure, sound and heat. The potential energy stored in explosive charge is result of chemical energy. Various chemicals undergo different explosive reaction and accordingly produce detonation.

Detonation is a process in which any explosive when blasts, there is generation of shock waves and velocity of the blast is faster than speed of sound in the medium. On other hand when explosive deflagrates, there is flame front that moves along the explosive column at a speed lower than speed if sound. During deflagration there are no shock waves generated and only gases are generated which creates pressure.

According to explosive rules the explosives and their accessories are classified into eight classes. These are:

A. Class—1 : Gunpowder
B. Class—2 : Nitrate mixtures (like ANFO, Aquadyne, Energel, GN-1, Godyne, Permadyne, Powerflow, Permaflow, Powerite, Superdyne, Supergel, Toeblast.)
C. Class—3 : Nitro compounds

D. Div:-1:Blasting gelatine, Special gelatine, O.C.G., Ajax-G, Viking-G, Soligex, etc. Div:-2: Gun cotton, PETN, TNT, etc,
E. Class—4 : Chlorate mixtures,
F. Class—5 : Fluminate,
G. Class—6 : Ammunition
    o Div:-1: Safety fuse, Igniter cord, Connectors, Electric lighters etc.
    o Div:-2: Cordtex, Detonating fuse, Plastic igniter cord, fuse, igniters, etc.
    o Div:-3: Detonators, Delay detonator, relays, etc.
H. Class—7 : Fireworks,
I. Class—8 : Liquid Oxygen Explosive (LOX)

Since the Bulk explosives are nitrate mixtures they come under class 2 explosives.

History

Initially Black powder was used as explosive for releasing rock from intact rock mass. History of use of Black powder dates back thousand years ago. Mining was very primitive. Black Powder is a mixture of Potassium Nitrate (75%), charcoal (15%) and sulphur (10%).

In 1865 Nitroglycerin based Dynamite was developed in Sweden and continued to be used till 1950's after which ANFO came in scene. Nitroglycerin is very unstable and handling needs lots of care. Dynamites were developed by Alfred Nobel who used diatomaceous earth (Kieselgurh) to absorb the NG and this could be easily transported to different places without much hazard of explosion.

Discovery of Ammonium Nitrate as an explosive was also because of an accident. There have been number of accidents

requii
signifi
All th
and C
Conte
SKO,
entire
mixer
Sorbi

Caust
solut
(sligh
Nitrit

Ra\

(SN

Amr
AN i:
fuel i
Amm
whei
whic
NH3

Tem
Crys
Cycli
32^0,
lead

during storage of AN, and though it is safe to transport AN, storing AN for long time can be very dangerous, because of its hygroscopic nature and inherent property of self caking and self heating. ANFO is today also used at many mines for blasting purpose though it has some limitations. Most commonly used explosive in modern time are Ammonium Nitrate based Site Mixed Slurry and Emulsion explosives.

ANFO

ANFO is mixture of AN (Ammonium Nitrate) and Fuel Oil. Ammonium nitrate is an oxidizer agent and fuel oil is reducing agent in explosive reaction (REDOX reaction).

In ANFO prills of AN is soaked in fuel oil namely High speed Diesel and mixed with small Amount of husks in a designed vehicle. This mixture is directly charged in to the blast hole. The loading should be carefully done so that the holes do not get chocked and slow tamping is to be done using wooden rod. This will help increase the charge density thus increasing energy concentration. The ANFO are very successful in dry holes but in watery holes the performance of ANFO is drastically reduced. This is because the AN is highly water absorbing and will get dissolved in water. This will reduce available energy for blasting purpose. Use of polyethylene bags can be used in case water in blast holes are less and if Polyethylene bags are resistant. Even pumping of water out from borehole was done for effective blasting using ANFO but these do not work in watery strata where water has huge infusion in holes.

thus becoming finer and finer eventually leading to fragmentation and disintegration of AN. The density of AN is 0.8 and due to cycling this this can increase upto 1.2 g/cc.

Also AN is hygroscopic, absorbs water and moisture, and then disintegrates leading to caking of AN

Due to these properties of AN, storage of AN for long time is not feasible. Also due to hygroscopic property, there is loss of AN from the bags into the storage area. To avoid losses, too much stocking of AN should be prohibited and First In First Out (FIFO) should be followed.

Calcium Nitrate (CN)

Chemical formula of Calcium Nitrate is Ca(NO3)2. It is also formed when Limestone (CaCO3) is treated with Nitric acid (Nitric Acid).

CN is a nitrate salt that can be used along with AN. The energy output from CN is less than AN.

CN also increases the density of Bulk thus increasing Bulk strength of explosive. Density of 95% CN solution is nearly 1.4 gm/cc.

Other Nitrate Salts and additives

Other nitrate salts like Sodium nitrate, Sodium perchlorate can be added to the SME. There are some additives that are added in oxidizer solution that help in gassing reaction. Additives like thiourea, salts of thiocynates acts as gassing accelerators. Also these additives help reduce generation of NO_x gases during gassing reactions. Another use of urea is that it stabilizes emulsions against thermal degradation in pyrite and sulphide ores.

Fuel Phase

In oxidation reduction (Redox) reaction there is exchange of Oxygen and reducing agent burns, thus releasing energy and same is the case with explosive reaction. Fuel phase is oxidized by gaining oxygen releasing the energy.

This phase is continuous phase and is primary made using fuels such as HSD, SKO, LDO, FO, Spent oils, waxes etc as per the availability. HSD provides the maximum energy.

FO or waxes or any other Fuel with high viscosity if added will increase the viscosity of the bulk explosive. In SMS guar gum is used to maintain viscosity.

Emulsifier and Co-emulsifier:

There are many emulsifier that can be used for stabilising SME explosives. Commonly used emulsifiers are Sorbitol mono Oleate (SMO) and Soya lecithin (SL).

It helps the aqueous solution of nitrate salts suspended in matrix of oil phase. Emulsifier has chemical structure such that it has one hydrophilic and one lipophilic region in the molecular structure. Hence water gets attached with hydrophilic group and oil gets attached with lipophilic portion of molecule, thus water and oil both are mixed together.

Sorbitol Mono Oleate is made from 2 components, Rice Bran Fatty Acid and Sorbitol blended together for 8 hours at a temperature of nearly 240^0 C. Since the temperature is sufficient to burn the oil, the chamber was filled using nitrogen gas. The Nitrogen gas cylinders were used for this purpose.

Sorbitol is made from Maize starch and chemically it is sugar alcohol. It is colourless or faint yellow syrupy liquid, with sweet taste. Sorbitol has a property that it is good humectants. Humectants are water stabilizer chemicals, i.e. they retard the loss of water by evaporation from oil-in-water emulsion and also inhibit absorption of water from atmosphere by any hygroscopic compounds.

Soy Lecithin: This is used as additional emulsifier in emulsion matrix. SL also has lipophilic component that gets attached to oil and hydrophilic portion that gets attached to water (Oxidizer solution). This provides stability and emulsion can be stored for longer duration.

If the quality of emulsifier is not upto the mark, there will be issue of emulsion stability and Nitrate solution will settle at bottom after breakdown of emulsion.

Parameters of oxidizer solution

pH: Desired range of pH of oxidizer solution should vary from 3-3.5 (slightly acidic). If oxidizer is formed has higher pH, then nitric acid is used and in case the solution has lower pH, then caustic soda is added to the solution. Role of pH is important as if pH is not maintained, then gassing of emulsion is very slow.

If pH is low (acidic), that will disintegrate thiourea/thiocyanates and gassing reaction may become slow and if pH is high (basic), these gassing accelerators do not work.

Crystallization Point: Crystallization point of any solution is the temperature at which first crystals of solute particles are formed. Oxidizer is a solution of AN, at high temperature (60-70^0 C) and this solution, temperature is reduced gradually, while continuously shaking oxidizer in flask. The temperature at which AN fine line crystals will become visible, that is crystallization point. The higher will be fudge point the higher concentration of AN in oxidizer solution.

Density: Density of oxidizer solution is 1.38-1.42 gm/cc. If product is made without CN, then this density will decrease.

Parameters of Fuel phase

Parameters that are monitored for Fuel phase are viscosity, GCV (Gross calorific value) flash point and density.

Parameters of Emulsion matrix

Viscosity, Droplet size, Emulsifier stability, VOD (in Field)

Testing Facility required for quality testing of raw material and final product.

AN: moisture content density, AN % purity

CN: CN content, density-1.6, visibility clear, Iron content .5% max,

Oxidizer: pH 3-4, density- 1.38-1.42 m3/kg and crystallization point

SMO: Sorbital mono oleate: viscosity, saponification test

Soya Lecithin: Viscosity, Lecithin content testing

FO: Density

HSD: Density

Fuel Phase: density, viscosity, Temp-70

Emulsion Matrix: Density and viscosity

Chemistry of Explosives

Reaction for explosive reaction in

3 NH4NO3 + CH2 - 3 N2 + 7 H2O + CO2 + high heat and shock energy

As can be seen that 14 gm of (1 mole) CH2 (Fuel) reacts with 240 gm (3 moles) of AN (5.5:94.5) ratio to produce Nitrogen gas, steam and Carbon dioxide. For the above explosive reaction to occur, high initial temperature is required.

If excess fuel is present then instead of CO_2, CO is formed. Formation of CO releases lower heat than formation of CO_2 as heat of formation of CO is higher than CO_2. So if excess fuel is present then lower energy is released as compared to balanced reaction.

On other hand, if oxidizer solution i.e. AN component is higher, then instead of N2, NOx will be produced. Production of NOx is harmful and also reduces the energy released in explosion reaction.

Emulsion Matrix prepared in plant is loaded into BMD vehicles and transferred to mine site where gassing agent is mixed and after formation of bubbles in the emulsion, only then mixture becomes explosive, hence it's called site mixed emulsion.

Slight negative oxygen balance is maintained for bulk explosives. This is done because CO formation will have lower energy losses as compared to generation of NO_x.

In the explosive reaction AN in present in emulsion as dispersed water and CH_2 is present as continuous oil phase. These are in ready contact with each other. These react when sufficient heat is provided to the bulk explosive. Reason why bubble formation is needed in emulsion explosive is it act as source of heat required for explosive reaction to initiate and continue (self sustain).

When booster is initiated then the bubbles present around are compressed instantaneously. As the volume of bubbles is reduced almost instantly (nearly adiabatic process), the temperature of these bubbles increases thus producing necessary heat for detonation of Explosive. The temperature of bubble rises from nearly from 50-60^0 C to 2000^0 C almost instantly. Hence these bubbles act as hotspots in the emulsion matrix. These hotspots provides necessary temperature for initiation of the explosive reaction, as well as propagation of reaction by further generating more hotspots and reaction becomes self propagating. Since these hotspots are present in the entire emulsion matrix hence these provide localized elevated temperature and hence chemical reactions happen at very high rate, which takes form of detonation. If hotspots are not present in some zone, then the reaction in that zone will not be detonative reaction and energy of explosive will be wasted. As detonation reaction proceeds there is release of huge amount of gases.

This concept of adiabatic compression is similar to that of ignition in diesel engine. Inside the engine, mixture of oil and gas is compressed via piston, thus producing very high compression causing burning of diesel. As the diesel burns, huge amount of gases will be released thus causing movement of piston thus converting it into useful work.

Nitrite salts are used for generation of chemical induced gas bubbles.

Nitrite ions reacts with ammonium ions to form gas bubble of N_2 gas and release water.

$NH_4^+ + NO_2^-$-------------- $N_2 + 2 H_2O$

If mixing of gassing agent is not uniform then there may be some portion in charge column where there is absence of gas bubbles. This will lead to fall in temperature and explosive reaction will rapidly cease to continue and will not stop instantly but travel to some length (this can be called area of sympathetic detonation where even a gap between two columns of explosive separated by some distance will detonate).

Various Properties of Explosive:

1: Density

Density is very critical property of explosive. Usually the density of SME emulsion is nearly 1.30+/- .05 gm /cc. Critical density is roughly 1.23 gm/cc. If density at any instance is higher than critical density then explosive will not initiate. This is because when gassing reaction takes place, it creates bubbles, which acts as hotspots. If due to some reason bubbles are not sufficiently formed, then detonation reaction will not propagate.

Density also signifies amount of energy per unit volume. If the density of explosive is reduced, then energy in given volume is reduced, so for softer or medium hard rock, density of explosive can be reduced, but in medium hard to hard rock, lower density emulsion will not suffice the purpose. Density of bulk explosive can be varied from 0.85 to 1.2 gm/cc for blasting purposes. For achieving lower density we can externally use air spaces, like using empty plastic bottles, polystyrene beads or some other techniques to further reduce density and but chemically by adding more of gassing agent should not be technique to increase bubbles, as that will lead to bigger bubble size, which will be unstable, and gases may leak out of emulsion.

Critical density is density at which sensitivity of explosive is reduced that even good primer is not sufficient to detonate the explosive column. Higher hydrostatic pressure can also desensitize an explosive by increasing its density.

Loading density: It is defined as amount of explosive required to cover 1 m length in hole. This is most important criterion in field operations. In general we consider that for 159mm diameter hole, 25 kg emulsion is required for 1 m of charged length. So loading density is 25kg/m.

Similarly for diameter of 259mm and loading density for general purpose is taken to be 65 kg/m. This can be derived using simple mathematical calculation:

Loading Density= Density of Explosive (g/cc)*0.000785*d^2(mm)

Depth (m)	Open Cup Density (g/cm³)							
	0.90	0.95	1.00	1.05	1.10	1.15	1.20	1.25
0	0.90	0.95	1.00	1.05	1.10	1.15	1.20	1.25
1	0.92	0.97	1.02	1.07	1.12	1.17	1.21	1.26
2	0.95	0.99	1.04	1.09	1.13	1.18	1.22	1.26
3	0.97	1.01	1.06	1.10	1.15	1.19	1.23	1.27
4	0.98	1.03	1.08	1.12	1.16	1.20	1.24	1.27
5	1.00	1.05	1.09	1.13	1.17	1.21	1.24	1.28
6	1.02	1.06	1.10	1.14	1.18	1.21	1.25	1.28
7	1.03	1.07	1.11	1.15	1.19	1.22	1.25	1.28
8	1.04	1.08	1.12	1.16	1.19	1.23	1.26	1.28
9	1.06	1.10	1.13	1.17	1.20	1.23	1.26	1.29
10	1.07	1.11	1.14	1.18	1.21	1.24	1.26	1.29
12	1.09	1.12	1.16	1.19	1.22	1.24	1.27	1.29
14	1.10	1.14	1.17	1.20	1.23	1.25	1.27	1.29
16	1.12	1.15	1.18	1.21	1.23	1.26	1.28	1.30
18	1.13	1.16	1.19	1.22	1.24	1.26	1.28	1.30
20	1.14	1.17	1.20	1.22	1.25	1.27	1.28	1.30
24	1.16	1.19	1.21	1.24	1.25	1.27	1.29	1.30
28	1.18	1.20	1.23	1.24	1.26	1.28	1.29	1.30
32	1.19	1.22	1.23	1.25	1.27	1.28	1.29	1.31
36	1.20	1.22	1.24	1.26	1.27	1.29	1.30	1.31
40	1.21	1.23	1.25	1.26	1.28	1.29	1.30	1.31
45	1.22	1.24	1.26	1.27	1.28	1.29	1.30	1.31
50	1.23	1.25	1.26	1.27	1.28	1.29	1.30	1.31
55	1.24	1.25	1.27	1.28	1.29	1.30	1.30	1.31
60	1.25	1.26	1.27	1.28	1.29	1.30	1.31	1.31
65	1.25	1.26	1.27	1.28	1.29	1.30	1.31	1.31
70	1.26	1.27	1.28	1.29	1.29	1.30	1.31	1.31
75	1.26	1.27	1.28	1.29	1.30	1.30	1.31	1.31
80	1.26	1.27	1.28	1.29	1.30	1.30	1.31	1.31

From the density chart it can be seen that as depth of hole increases the density of explosive at that depth also increases because of self weight and hydrostatic pressure in the borehole. If the cup density (that is density at atmospheric pressure) is 1.1 gm/cc yet this SME bulk will not explode if depth of hole is above 30m as dead press density of explosive is reached. In deep hole charging there is provision for addition of LDAN Prills (low density AN Prill). These prills prevent bulk from reaching dead press as these prills have air trapped inside, that acts as hotspots in the bulk. Hence doping straight emulsion with AN prills work very well in deep hole benches.

Formation of toe is a major issue in deep hole blasting, and with introduction of bottom initiating .systems like Electronic detonator or NONEL, this issue has been reduced significantly. Earlier the blasting was done using detonating fuse, and major drawback of DF is it is top initiating system and also desensitise emulsion that is in contact with the DF.

2: Velocity of Detonation (VOD)

The detonation velocity is a measure of speed at which the detonation wave travels through column of explosives.

When an explosive detonates, then the explosive detonates and produces a huge amount of gases at high temperature and pressure along with shock energy release. The plane in the explosive column at which the explosive detonates that plane is called CJ Plane. As the explosive detonated this CJ Plane moves and velocity of movement of this reaction plane in explosive is called Velocity of detonation. VOD of Bulk explosives range from 3500-4500 m/s.

As a general rule, the higher the velocity of detonation the greater is shattering effect

It has been observed that even though many times the VOD of explosive is lower yet still fragmentation is good. This can be justified as VOD is actually measure of Shock energy, and as we know that shock energy is not primary factor controlling fragmentation rather 85% work is done by gas energy. Detonation velocity is important consideration for applications outside a borehole, such as plaster shooting and mud capping but it plays significantly less importance if explosives are used in the borehole.

For Selection of explosive for a given rock type, VOD is very important criterion. Impedance of rock and of explosive should match for good result. Impedance of rock is product of seismic wave velocity in rock and density of rock and impedance of explosive is product of VOD of explosive and density of explosive. So for hard rock with high density and uniform strata, explosive with higher density and VOD is required for optimum result.

Various factors affect VOD of explosive are mentioned below:

Explosive Type:
VOD of slurry and emulsion explosive is 4000+/- 500 m/s, but can change as per the requirement of the customer. Density range of Emulsion is lower as compared to slurry. VOD of ANFO is slightly higher than SME emulsion that is manufactured.

Charge Diameter:
As the charge diameter increases the VOD increases and becomes constant after reaching its thermodynamic VOD (Ideal VOD). For SME, it is found that ideal VOD is achieved in nearly 300mm diameter hole. Also, if diameter of confining charge is

reduced, then VOD decreases and after critical diameter, the charge will not detonate instead it will deflagrate. This concept can be seen in Detonating Fuse and NONEL. In detonating Fuse, the PETN charge is above its critical diameter, hence DF when initiated detonates and produces shock and noise. On the other hand, NONEL has inner tube diameter, where PETN charge is stored, is less than critical diameter, so there is no shock or noise when NONEL tube is initiated.

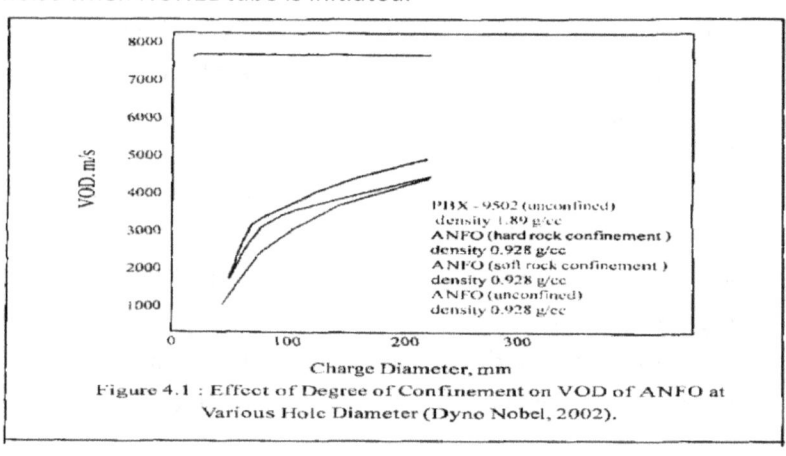

Figure 4.1 : Effect of Degree of Confinement on VOD of ANFO at Various Hole Diameter (Dyno Nobel, 2002).

Explosive formulation:

VOD of explosive is dependent on the ingredients of oxidizer and fuel phase present in emulsion. Oxidizer solution prepared using pure Ammonium Nitrate has maximum VOD and if there is any substitution, using calcium or Sodium Nitrates, VOD is substantially reduced.

Addition of Aluminium powder, increases the VOD of explosive. Sulphur addition increases the stability of Slurry explosive and other additives like Thio-Urea are added and impacts VOD of Emulsion.

Effect of Sensitizing Agent:

In Slurry and emulsion explosive, gas bubbles acts as sensitizer. Size of gas bubble has impact on VOD. If size of gas bubble increases, VOD will decrease and vice versa. VOD decreases as bubble size increases.

Effect of Density:

VOD increases as density is increased but to a maximum limit and then it falls.

Density range for maximum VOD for emulsion explosive is 1.1-1.2 g/cc and as density of product decreases, VOD will decrease. Emulsion explosive will not detonate if density is more than 1.24 g/cc which is also called dead press density. Initial density of explosive is 1.30-1.35 g/cc and this density is reduced using chemical gassing, which leads to generation of gas bubbles. The final density depends on the number and size of gas bubbles generated. Also in deep holes the density of explosive increases due to hydrostatic pressure. A density chart is used for charging the hole in deep holes charging.

Important thing to note is that density that is achieved by chemical gassing is being considered. If additional gas bubble is inserted using beads of empty bottles or other gas bubbles in small amount, then VOD of the explosive should not change. Effect of sensitizing agent and density are almost the same cases, for VOD variations.

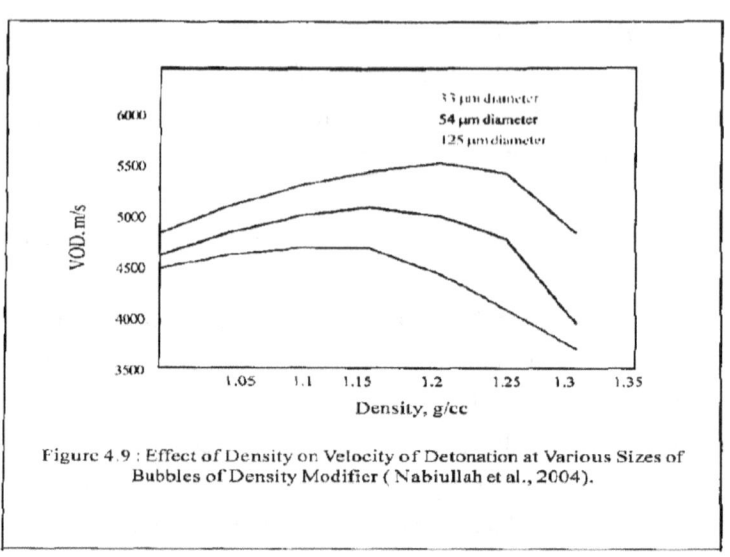

Figure 4.9 : Effect of Density on Velocity of Detonation at Various Sizes of Bubbles of Density Modifier (Nabiullah et al., 2004).

Temperature: VOD of explosive depends on the temperature of surrounding rock. Hot holes having temperatures above 80^0 C are not charged because this may cause premature explosion. It has also been observed that detonators detonate if put at 100^0 C for 2 hours or more. For emulsion explosives the VOD increases as the temperature increases till nearly 75^0 C and then VOD starts falling. Emulsions have huge impact if temperature falls. As emulsion has highest component as AN dissolved is water and solubility decreases as temperature falls, hence AN will crystallize out and effect blasting performance.

Priming:

VOD of bulk in hole depends on the amount and type of primer used. In current scenario, emulsion booster is being used. Primary issue with them is they do not have sufficient VOD and energy for initiating emulsion explosive to reach steady state VOD, hence many times fragmentation is not up to the mark. 0.2% cast booster as primer is said to be sufficient for initiation of Bulk emulsion (as per study conducted by NIRM 2001).

Doping of SME emulsion increases the VOD of explosive.

VOD measurement of SME explosive: D- Autrich Method
D-autrich method of VOD Calculation

D- Autrich Method

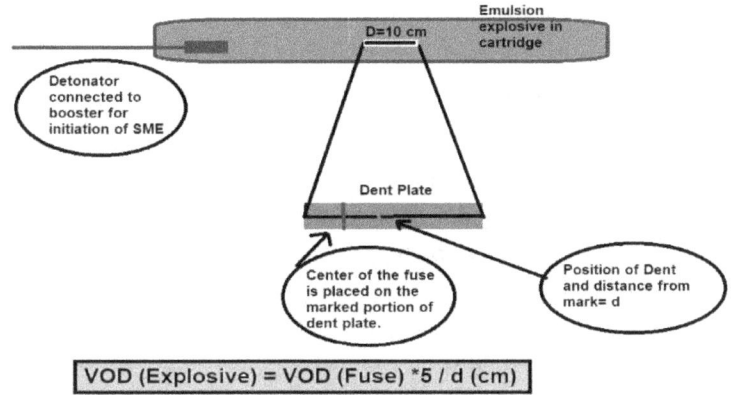

VOD (Explosive) = VOD (Fuse) *5 / d (cm)

Requirements:

Dent plate, VOD of Fuse should be tested using VOD meter

Detonator and cast booster for initiation of explosive

Emulsion Explosive is put in plastic or cartridge form with diameter of 83 mm. Initiation of booster is done via NONEL or electric detonator. A mark is made on dent plate on one side. A fuse of nearly 2m or more is taken and midpoint of fuse is placed on the mark made on dent plate and fuse is tightly tied to plate via tape. One end of hanging fuse is longer and this end is put inside cartridge (perpendicularly) near booster side (at least 25 cm away from booster), and shorter end is put at distance D(=10 cm)form the first end.

Calculation:

Concept is simply time consideration

Uniform mark will be formed along the plate due to burning of Fuse, but dent is formed in plate at the point where burning from both fuse will meet. Let this dent be at distance "d" from mark on plate.

If length of fuse is L, then time to cover L/2+d from fuse is same as time to cover D(=10 cm)+L/2-d

$(L/2+d)/VOD_{fuse}=(D=10)/VOD_{expl} +(L/2-d)/VOD_{fuse}$

Equating the above we get

$VOD_{expl} = VOD_{fuse} *5/d$(in cms) in m/s

Now we have number of VOD meters that can measure VOD of explosive directly.

In the hole VOD can also be done using VOD meters which are based on resistance wire continuous VOD system. In this continuous current is passed to the wire and as the wire gets shortened, there is drop in Voltage at the rate same as changing length of wire. So a correlation is done for measuring VOD. Our company is planning on purchase of VOD meter for in-the-hole VOD measurement.

Steady state VOD:

The VOD attained and that is sustained after run up length for an explosive charge column is called Steady State VOD. There are a number of factors that determine the steady state VOD for the explosive charge.

1. Diameter of charge column
2. Quality of explosive
3. Confinement
4. Density

3. Critical Diameter

Minimum diameter at which explosive once initiated will support itself in column. Higher the diameter of confinement space higher is the VOD until steady state VOD is achieved. For

SME bulk critical diameter is nearly 75 mm. If diameter of hole is less than critical diameter, then explosive will not detonate rather it will deflagrate.

The same concept of critical diameter is used while designing NONEL. Inner lining of NONEL and detonating fuse both have PETN, yet when detonating fuse (inner diameter is 4-5mm) nearly is initiated it detonates at a VOD of 7000m/s where as when NONEL (inner diameter is 1 mm) is initiated, PETN inside deflagrates at the VOD of 2000m/s and there is no noise or shock when NONEL is initiated.

4: DETONATION AND BOREHOLE PRESSURE

Detonation pressure is a function of the detonation velocity and density of an explosive. The detonation pressure is more dependent on detonation velocity than specific gravity. A high detonation pressure is necessary when blasting hard, dense rock. In softer rock, a lower pressure is sufficient. Detonation pressures of explosives range from 10 to over 140 Kilobars.

Detonation Pressure can be approximated as

$P = 2.5*D*V^2/1000000$ (in Kilobars)

Where, D= density (g/cc), V= velocity(m/s)

5: Water Resistance

The ability of explosive to withstand water penetration is termed as water resistance. It is generally expressed as number of hours a product may be submerged is static water and still be detonated reliably. Explosives are rated on water resistance basis like class 1 water resistance suggest that explosive under water for 72 hrs has no detrimental effect on property, class 2 for 48 hrs, Class 3 for 24 hrs and class 4 for 12 hours.

ANFO has very low water resistance

Also low density products can easily become separated by floating in dirty water. Static water at low pressure will not affect explosive as quickly as dynamic water especially at high pressure. Slurry and emulsion have good water resistance.

The issue of water resistance becomes acute in low viscosity product. Though SME explosive is highly resistant to water but if viscosity of SME is low, there are chances of penetration of water in SME and doesn't gives desired blasting result.

Another issue due to viscosity is that rate of gassing reaction in SME is very fast. In such cases blasting performance is reduced (especially in hard benches).

If water in strata is flowing (in motion) then SME will be washed out and blasting will not be very good.

6: Sensitivity:

Sensitivity is measure of ease of initiation.

Cap Sensitivity: It is measure of minimum energy required for initiation of detonation of explosive. No. 8 test blasting cap is used for this purpose. No 8 blasting cap consist of 2 g mixture of 80% Mercury Fulminate and 20% Potassium Chlorate. ANFO, slurry and SME explosives are non cap sensitive and are initiated using cap sensitive explosive primer.

With the increasing use of emulsion booster as a primer, there are cases when these boosters could not initiate emulsion explosive. The reason could be that if quality of emulsion booster is not up to mark, then it does not provide sufficient energy to initiate bulk emulsion.

7: Sensitiveness:

This is the characteristic of explosive that determines the ability of explosive to propagate detonation throughout the length of explosive column and controls the minimum diameter. This also measures the maximum distance by which a charge is separated such that primed donor cartridge and un-primed receptor cartridge can be initiated. This property is called Gap sensitivity of the explosive.

8: Fumes

Gases releasing after detonation of explosive are carbon dioxide, Nitrogen and Water Vapor which is non toxic. However if there is oxygen imbalance there will be formation of Nitrogen oxides and carbon monoxide, which are toxic and NOx are yellow fumes.

Oxygen balance

If there is insufficient oxygen (negative Oxygen balance) in the explosive that is Fuel % is higher, then there will be formation of Carbon Monoxide. Formation of Carbon Monoxide has higher heat of formation so the energy released by explosive will be reduced.

If there is excess of oxygen (Positive Oxygen Balance) in the explosive, that is oxidizer % is higher, then there will be formation of Nitrogen Oxides, which again has higher heat of formation leading to reduction in energy of explosive. For optimum oxygen balance, % of Oxidizer to Fuel ratio must be 94:6

Energy released in explosive reaction when CO is formed is more as compared to energy released when NOx are formed, so even oxygen balance is done with slightly higher % of fuel. For optimum AN: Fuel ratio is 94.5:5.5, yet for normal purposes composition is made at 94:6 ratio (slightly negative oxygen balance).

Presence of fumes doesn't merely depend on Oxygen imbalance, rather there are many factors that are responsible for fumes generation in blast.

Presence of Water, Charging of deep holes, presence of some impurities in raw material like sulphur content in emulsifier or fuel, iron content in CN or other impurities, inadequate priming are few causes of fumes generation.

If there is presence of reactive rock strata or some mineral content in strata that reacts with emulsion, it may cause fumes.

9: Strength

Strength of explosive is the energy that explosive releases upon detonation. This can be done using number of method like ballistic mortar, underwater bubble, catering and strain pulse test.

There is not a single test that can directly determine the strength of explosive. There is concept of theoretical energy and Expansion work done by gases, which calculated can indicate the energy released by explosive. Yet the strength is not simply amount of energy released rather, energy utilized and rate of release of energy. For example VOD of explosive gives the rate of shock waves moving in explosive column, but it doesn't give any idea on gas energy released and total energy utilized.

BMD vehicle:

BMD vehicles are used for transporting bulk emulsion to Mines from plant. They are installed with number of positive displacement screw pumps, and monitors for temperature, pressure, product flow, rotor RPM and counter (for amount of explosive charged in holes) on rotation of product pump using proximity sensor.

Usually a proximity sensor is placed in product pump and this sensor gives count of RPM of rotor, hence the flow rate of emulsion on which concept counter system works. This is the reason that if viscosity of product changes, then same counter will deliver less product if viscosity increases and vice versa.

BMD Vehicles are mainly of 2 types: Trucks with doping facility and trucks for straight emulsion.

Trucks for Doped Emulsion:

They have chambers namely for following purposes

1: Emulsion Chamber: For carrying emulsion matrix. This bin covers the maximum volume of the BMD vehicle. Capacity of BMD depends on the size of this bin.

2: Water Chamber: Water is used as lubricant for free flow of emulsion when it passes through delivery hose pipe, and also for flushing of explosive. If water is not available, that will lead to chocking of hose pipe and pressure in the system will increase and due to that BMD will automatically trip.

3: Gassing agent Chamber: At our plant we use Sodium Nitrite as gassing agent for explosive. A solution of Sodium nitrite in water in filled in this chamber.

4: Acetic acid tank: Acetic agent acts as catalyst for gassing reaction and may be added in this chamber. This may be required when temperature of emulsion matrix fall and rate of gassing is slow.

These 4 chambers are available in straight emulsion trucks also. The only difference in doped truck is addition of chamber for Low density ammonium nitrate prills (LDAN Prill).

Beside advanced features in BS VI engines related to electronic circuits and control of pollution, presence of governor for speed control.

In BS IV and BS VI engines, there is introduction of SCR system (Selective Catalytic Reduction). During normal running of diesel engine there is formation of NOx. NOx is highly greenhouse gas and measures are taken for conversion of NOx into N2 which is easily done when NOx is heated in urea water. Since the exhaust gases are already at high temperature, so in SCR, droplets of urea water are continuously spilled and exhaust gases are passed, which causes reduction of NOx into N2.

BMD is primarily a hydraulic system installed on truck for delivery of emulsion along with other materials necessary for blasting.

Safety Features in BMD
Intrinsic Safety

Though emulsion is highly resistant to shock and heat and friction in terms of initiation yet sufficient safety features are incorporated in BMD vehicles.

Intrinsic safety features-

The whole compartment of BMD is made of Stainless steel, and internally each compartment is lined with steel liners, which prevent mixing of components present in different compartments inside the vehicle.

The internal parts of all positive displacement pumps are made of rubber. This provision is made so

that if any foreign particles fall in any compartment of BMD, they do not generate heat due to friction. These foreign particles will damage the internal lining and there will be no flow of explosive from that pump.

Hose used for delivering the bulk explosive into the hole are antistatic. These are lined with steel wires so that if any stray charge is accumulated in emulsion or BMD these can provide grounding for these stray charges and minimizing chances of initiation of detonators in holes due to these stray charges.

Rupture Disc; Temperature shut off valve; pressure shut off valve;

Externally we have parking breaks, fire extinguishers; earthing chain, speed governor, first aid box;

Presence of Pressure valves: Pressure valves detect pressure in system, trips the BMD if pressure exceeds the specified value. This pressure transducer is attached in progressive cavity pump (Product Pump) for discharging the emulsion matrix into the delivery hose.

Presence of temperature sensors: Temperature in emulsion may rise due to friction caused in system if any foreign body (Iron particle, rocks). Though Emulsion matrix is resistant to heat, yet it is it is very important to prevent rise in temperature. To prevent any issue related to this, thermal sensors are installed in BMD vehicle and when temperature rises above 80^0 C, entire BMD system will trip down.

Pressure rupture disc: Rupture Discs are installed in new BMD vehicles which prevents increase in pressure in delivery hose. This disc is installed in product pump and bursts if there is increase in pressure in delivery Hose.

Earth chain for dissipation of stray current, if any, accumulated in the BMD, emulsion. This prevents chances of initiation of detonator or generation of spark due to these stray charges which can initiate detonators.

There are chances of generation of static charge whenever there is friction between two bodies. So during pumping of Explosive there are chances of generation of static charges. If these static charges are not dissipated then it may generate some flash, stray current that can initiate bulk explosive. These charges need to be dissipated and hence earth chains and provision of anti static hose is present in BMD.

Separation of each chamber is done using internal lining of stainless steel.

Delivery Hoses of BMD is provided with steel wire reinforcement. This not only provides strength to hose but also dissipated any stray electric charges that may be generated because of friction as emulsion flows through the delivery hose. Diameter of delivery hoses are less than critical diameter of bulk explosive, so even in case of blasting, the issue of blasting of emulsion in hose will not take place.

Extrinsic Safety

A: Fire Extinguisher: Fire extinguishers are available in BMD vehicle to prevent any incident of fire which may be produced due to any electrical fault or other reasons. Extinguisher used is dry CO_2 type and ABC type.

B: Earth Chains: Earth chains are provided in BMD vehicle and when ever charging of blast holes are being done, BMD is grounded, as to dissipate any static charge that may be accumulated during charging process. Static charge if not dissipated may cause spark and can lead to accident.

C: Parking Brake:

D: Wooden Block for parking:

E: First Aid Box

F: Reverse Horn

G: Safety Belts

Multiple Choice Questions

1: What is detonation?

A: The rate of burning is higher than speed of sound	C: For effective blasting in medium hard and hard rocks, explosives must detonate for proper blast result
B: A substance when detonates produces shock waves, along with heat, gases and sound	D: All of these

Ans: D

2: What is deflagration?

A: The rate of burning is lower than speed of sound	C: For effective blasting in medium hard and hard rocks, explosives must deflagrate for proper blast result

B: A substance when deflagrates produces shock waves, along with heat, gases and sound	D: All of these

Ans: A

3: Chemical formula of AN

A: NH_4NO_3	C: $Al(NO_3)_3$
B: NH_4NO_2	D: $Al(NO_2)_3$

Ans: A

4: Expand PETN

A: Poly ethane tri nitrate	C: Poly ethylene Tetra Nitrate
B: Penta Erythritol Tetra Nitrate	D: Penta ethene tetra nitramine

Ans: B

5: State of Nitro glycerine at room temperature

A: Solid	C: Gas
B: Liquid	D: Plasma

Ans: B

6: Expand SME

A: Site Mixed Explosive	C: Site mixed slurry
B: Site mixed Emulsion	D: None of the above

Ans: B

7: What is emulsion

A: Mixture of 2 miscible liquids	C: A liquid with suspended air bubbles
B: Mixture of 2 immiscible liquids	D: None of the above

ANS: B

8: Components of SME emulsion

A: Oxidizer Phase	C: Both of the above
B: Fuel Phase	D: None of the above

Ans: C

9: What is explosive

A: Any substance that burn rapidly and produces huge energy in form of shock, gas pressure, sound and heat	C: Substances that detonates
B: Substances that deflagrates	D: All of these

Ans: D

10. What is incorrect about SME explosives

A: The raw materials are mixed and final product is prepared at mines site.	C: These are brought to mines site using special vehicles called explosive van
B: It is prepared using two immiscible liquids	D: These are used in variety of rock types and there energy can be varied

Ans: C

11. Black Powder is mixture of

A: Potassium nitrate, charcoal and Sulphur	C: Ammonium Nitrate and Fuel oil
B: Calcium Chlorate, Sulphur and wood	D: PETN and Fuel oil

Ans: A

12: Which is true about SME Explosive

A: Water Phase is dispersed in continuous Oil Phase	C: The two immiscible liquids are mixed together using emulsifiers for

	stability of emulsion
B: Ammonium Nitrate is in Water Phase and Fuel Oil in Oil Phase	D: All of these

Ans: D

13: Components of Oxidizer solution

A: Ammonium Nitrate	C: Catalyst for gassing reaction
B: Other nitrate salts	D: All the above

Ans: D

14. Component of Fuel phase

A: Liquid Fuels	C: Acid or Base for maintaining pH of solution
B: Emulsifiers	D: Water

Ans: A and B

15. pH of Oxidizer solution should be

A: Slightly acidic	C: highly acidic
B: Slightly alkaline	D: Highly alkaline

Ans: A (pH range is nearly 3.5-4)

16. What is incorrect about Ammonium Nitrate

A: The crystalline structure changes with change in temperature	C: AN undergoes fragmentation and caking that leads to decrease in its density as storage time increases
B: It is hygroscopic substance	D: If stored for long there is loss of AN especially in rainy season.

Ans: C, AN used in explosive manufacturing is in form of prills and have lower density. Due to changing crystalline structure, these AN gets are first break down into fines and then due to hygroscopic nature forms cake. These turn into hard rock of AN with density higher than initially used AN.

17. Properties of Calcium Nitrate

A: it releases NO_3^- ions that assist in explosive reactions	C: It has density of nearly 1.4 g/cc
B: it is formed on reaction of Limestone with Nitric acid	D: All of these

Ans: D

18. Crystallization point is

A: Minimum Temperature at which there is formation of crystal of solute in a solution	C: Maximum time taken by solution during cooling for formation of crystal of solute in non saturated solution
B: Maximum temperature at which there is formation of crystal of solute in a non saturated solution	D: Minimum time taken by solution during cooling for formation of crystal of solute in non saturated solution

Ans: B

19. What is Ideal ratio of AN and fuel oil in ANFO

A: 94:6	C: 16:84
B: 84:16	D: 6:94

Ans: A

20. What is effect of increasing Fuel oil percentage in explosive reaction

A: Formation of some quantity of Carbon Monoxide	C: Formation of NO_x
B: Reduced output of	D: Reduced energy output

explosive as heat of formation of CO is more than CO_2	of the explosive as heat of formation of NO_x is more than CO_2

Ans: A and B

21. What is effect of decreasing Fuel oil percentage in explosive reaction

A: Formation of some quantity of Carbon Monoxide	C: Formation of NO_x
B: Reduced output of explosive as heat of formation of CO is more than CO_2	D: Reduced energy output of the explosive as heat of formation of NO_x is more than CO_2

Ans: C and D

22. What are the properties of emulsion matrix to be considered

A: Viscosity	C: Saponification test
B: Density	D: pH

Ans: A, B and D

23. What should be oxygen balance for the explosives:

A: Slightly oxygen negative	C: Oxygen balanced
B: Slightly oxygen Positive	D: Independent of oxygen balance

Ans: A; Although the primary aim is to be oxygen balanced, yet emulsions are slightly negative so that chances of NO_x generation is minimized. This is because energy output is very significantly reduced if NO_x is formed.

24. Importance of gas bubbles in emulsion

A: They acts as hotspots	C: They help in increasing

	temperature because of isothermal compression taking place
B: They propagate explosive reaction by providing elevated temperature	D: The temperature of these hotspots rises to 150^0 C, thus providing heat in the explosive

Ans: A and B

25. Which of the following chemical can be used as gassing agent for emulsion explosive?

A: Calcium Nitrate	C: Ammonium Nitrate
B: Sodium Nitrite	D: Potassium Perchlorate

Ans: B

26. What is the gas formed during gassing of Emulsion

A: N_2	C: SO_x
B: NO_x	D: CO_2

Ans: A

27. What is critical density of explosive.

A: The maximum density at which bulk explosive will detonate	C: At critical density the hotspots are just sufficient to propagate an explosive reaction
B: Minimum density at which bulk explosive will detonate	D: For Emulsion Bulk explosive, this value is nearly 1.00 g/cc

Ans: A

28. Density of explosive in hole depends on

1: Gassing Agent mixed properly

2: Depth of Blast Hole

Options

| A: 1 is correct | C: Both are correct |
| B: 2 is correct | D: None are correct |

29. Loading density is

1: Explosive charge required per m of blast hole

2: Depends on the diameter of the hole

Options

| A: 1 is correct | C: Both are correct |
| B: 2 is correct | D: None are correct |

Ans: C

30. Role of LDAN prills in emulsion explosive

| A: Increases the strength of the explosive | C: Prevents explosive from getting dead pressed in case of deep hole blasting |
| B: Acts as hotspots in the explosive charge column | D: All of these |

Ans: D

31. Dead press density is

1: When the cup density of explosive is lower than critical density but in hole density increases above critical density of explosive

2: The reason for dead pressing of explosive charge is due to increased hydrodynamic pressure exerted on the charge column.

Options

| A: 1 is correct and 2 is the correct and 2 is correct reason for 1 | C: 1 is incorrect |
| B: 1 is correct and 2 is the correct and 2 is wrong reason for 1 | D: 2 is incorrect |

32. Density of explosive in charge column can be reduced by

1: Increasing the gassing agent to certain limit

2: Externally adding air filled materials in the charge column

Options

A: 1 is correct	C: Both are correct
B: 2 is correct	D: None are correct

Ans: C

33. What is effect of Increasing density of explosive in the charge column

1: Higher density product has higher VOD and concentrated energy, thus suitable for hard rocks

2: Lower density product has lower VOD and thus suitable for soft, fragile strata

Options

A: 1 is correct and 2 is the correct and 2 is correct reason for 1	C: 1 is incorrect
B: 1 is correct and 2 is the correct and 2 is wrong reason for 1	D: 2 is incorrect

ANS: B

34. VOD stands for

A: Velocity of deflagration	C: Volume of detonation
B: Velocity of detonation	D: Volume of Deflagration

Ans: B

35. VOD is

1: Measure of velocity at which the reaction propagates in the explosive charge column

2: Rate of movement of CJ plane in the charge column

Options

A: 1 is correct	C: Both are correct
B: 2 is correct	D: None are correct

Ans: C

36. Higher VOD is indicative of

1: Higher shock energy of explosive

2: Lower Gas volume generation in explosive

Options

A: 1 is correct	C: Both are correct
B: 2 is correct	D: None are correct

Ans: A

37. Impedance of explosive depends upon

1: Density of explosive

2: Velocity of Detonation

Options

A: 1 only	C: both
B: 2 only	D: None

Ans: B

38. Impedance of rock is given by

1: Density of Explosive used

2: Speed of propagation of waves through the rocks

Options

A: 1 is correct	C: Both are correct
B: 2 is correct	D: None are correct

Ans: B, depends on density of rock type

39. What is effect of charge diameter on VOD

1: VOD Increases to steady state, with increasing diameter	3: No relation

2: VOD Decreases to steady state, with increasing diameter	4: First increases, and then decreases with increasing diameter.

Ans: 1

40. Which graph depicts effect of charge diameter on VOD (X axis Charge Diameter and Y axis VOD)

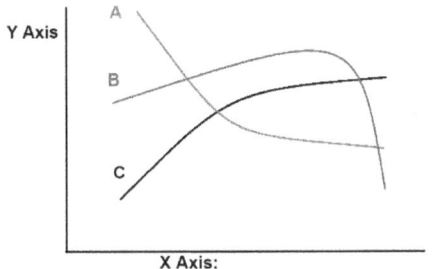

Option: A, B, C or None

Ans: C

41. Which graph depicts effect of density of explosive on VOD (X axis Charge Diameter and Y axis VOD)

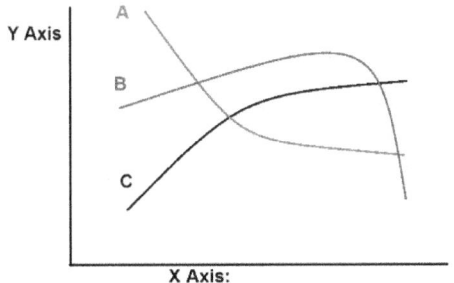

Option: A, B, C or None

Ans: B

42. What is critical Diameter for explosives?

A: Minimum Diameter at which explosive sustains its detonation.	C: Minimum diameter at which explosive will burn (deflagrate)
B: Maximum Diameter at which explosive sustains its detonation	D: NO such term

Ans: A

43. Density of explosive in bottom of charge column

1: Increases due to hydrostatic pressure

2: Decreases due to hydrostatic pressure

3: Remain unchanged

Which statement is correct

Ans: 1

44. According to DGMS, what is the maximum temperature of hole that can be charged with explosive.

1: 40⁰ C	3: 80⁰ C
2: 60⁰ C	4: 100⁰ C

Ans: C

45. Type of Primer used in blasting determines the VOD of explosive column in what way

A: No relation of Primer on VOD of explosive charge column	C: VOD of primer is high, then VOD of explosive charge will reduce
B: VOD of primer is high, then the VOD of explosive charge will increase	D: None of the above

Ans: B

46. What are Toxic gases released during blasting

A: CO_2	C: NO_x
B: N_2	D: CO

Ans: C & D

47. What could be probable reason of generation of fumes in blast

A: Improper ratio of Fuel and Oxidizer	C: Reactive strata
B: Leaching of explosive due to strata water	D: Presence of impurities in raw material used for manufacturing emulsion explosive

Ans: A, B, C and D

48. ANFO has

1: High water resistance

2: It gets easily dissolved in water present in strata

Which of the following statement is right?

A: 1 & 2 are correct and 2 is correct explanation of 1	C: 1 is correct and 2 is wrong
B: Both correct but 2 is not explanation of 1	D: 2 is correct and 1 is wrong

Ans: D

49. Emulsion explosive have

1: High water resistance only in case of static water present in strata

2: Flowing water will wash out Emulsion explosive from charge column

Which of the following statement is right?

A: 1 & 2 are correct and 2 is correct explanation of 1	C: 1 is correct and 2 is wrong
B: Both correct but 2 is not	D: 2 is correct and 1 is

explanation of 1	wrong

Ans: B

50. How much time does Class I water resistant explosives remain in water with no degradation in quality?

A: Minimum 72 hours	C: Minimum 50 hours
B: Minimum 100 hours	D: Minimum 48 hours

Ans: A

51. What is composition of Number 8 Blasting cap

A: 2 g mixture of 50% Mercury Fulminate and 50% Potassium Chlorate	C: 10 g mixture of 50% Mercury Fulminate and 50% Potassium Chlorate
B: 2 g mixture of 80% Mercury Fulminate and 20% Potassium Chlorate	D: 10 g mixture of 80% Mercury Fulminate and 20% Potassium Chlorate

Ans: B

52. What are BMD vehicle used for?

A: Manufacturing of Bulk Explosive at mine site	C: These are special vehicles that are mobile manufacturing units for bulk explosive
B: Transportation of non explosive materials of Emulsion explosives to mines site	D: All of the above

And: D

53. What type of fire extinguisher are available in BMD vehicles?

A: ABC type dry chemical fire extinguisher	C: D type fire extinguisher
B: CO_2 type fire extinguisher	D: All of the above

Ans: A and B

Chapter 3: Properties of Rock Mass

Rock Strength:

Higher the compressive and tensile strength of rock, higher is the strength of explosive required for proper fragmentation. For a given energy by explosive, the size of fragment varies considerably with change in rock mass properties. Uniaxial tensile strength plays important role in fragmentation. The actual fragmentation of rock takes place in tension (when the shock waves return after reflection from free face), as tensile strength of rock is very low as compared to compressive strength.

Brittle index is the ratio of compressive strength to tensile strength. This can vary from 10-100 for varying rock type. More is the Brittle index more is the rock friable in tension and since the actual fracturing of rock takes place in tension, hence more brittle rock give better blast fragmentation. So one should consider brittle index while blast design. For a brittle rock, one can increase blast pattern (i. e. Burden and spacing).

Rock Density

In general, the density of rock usually related with strength of rock. Higher the density of rock, higher will be strength of rock. This case cannot be correct all the time, as in case of fractured strata. Competency of rock type is also important for selection of explosive beside impedance. Impedance of rock is product of seismic wave velocity in rock and density of rock and impedance of explosive is product of VOD of explosive and density of explosive. So for hard rock with high density and uniform strata, explosive with higher density and VOD is required for optimum result.

Elastic Properties of rock

Elastic property of rock is an important factor that help in determining the selection of explosive used in the rock. It gives the ability of rock to absorb shock waves that are generated

IF Plastic nature of rock is dominant, then there will be absorption of shock waves and that will get easily attenuated. So tensile waves generated will be less and thus fragmentation due to spalling will decrease. Most of the work will be done by gas pressure but still fragmentation will be better.

If elastic nature of rock is dominant, the rock will readily propagate shock waves which on reflection from free face will generate tensile waves that will create the cracks along the rock mass. Usually elastic rocks have higher strength, thus higher energy product may be required.

The worst case can be that in matrix of plastic rocks, there are elastic rocks present in the suspended state. In this scenario, the suspended rocks will not break as they will not get tensile waves for the breaking of rock and gas pressure will only be able to displace these rocks and no breaking of boulders will take place. This is very common situation in top benches where the hard pre formed boulder is suspended in softer sandstone or clayey sandstone. In this situation there are maximum chances of boulder generation if drilling is not done in the preformed boulder itself.

Poisson's Ratio:

Poisson's Ratio is the negative of the ratio of transverse strain to axial strain.

For sandstone or usually other rocks Poisson's ratio is nearly 0.2. Cork has nearly 0 and steel has nearly 0.5 Poisson's ratio value. For the higher plastic rock value of Poisson's ratio is tending to 0 and higher the Poisson's ratio higher is the elastic nature of rock.

Wave properties of rocks

The wave properties represent the generation and velocity of shock waves that are produced in rocks during blasting or any seismic event. The higher the velocity of wave propagation in the rock, higher will be impedance of the rock strata. Selection of explosive is dependent on the impedance of the rock strata. For higher Wave velocity for a given rock, the required Velocity of detonation for given rock will be higher.

Presence of Discontinuity- fault planes

Presence of discontinuity leads to generation of free face, from where the compressive waves will reflect back and will not cover the entire blast face and lead to improper fragmentation and generation of boulders. Also since this fault will be least resistance path, gases generated post blast will move out from this fault plane. This has been usually found that in fractured areas or rock mass with number of faults, blast performance is usually not up to the requirements.

Presence of cavity

If any Cavity is present in the rock mass and blast hole passes through it, measures should be taken that the charge column should not fall in this region of cavity. In this scenario decking is also not a solution as all the material would go into cavity. In this case we lower a cut bag filled with sand using rope or DF and place it in such a way that it lies just above the cavity and provides a false bottom an which top charge can be placed.

Due to cavity, the fragmentation is poor as the gas generated post blast will have a easy way out, thus reducing the heaving effect of gas energy. Also it will provide false free face because of which the shock wave energy is also not utilized properly.

Variation in strata in a bench:

Variation in strata is very common and one of a major reason of blast failure at many patches. Reason of Blast failure is due to variation in strata, the partition layer or transition layer have weakness and provides a path of low resistance to gases generated in blasting, thus reducing the gas energy available for heaving. To cater this problem, one must place some deck charge at places where there is soft strata.

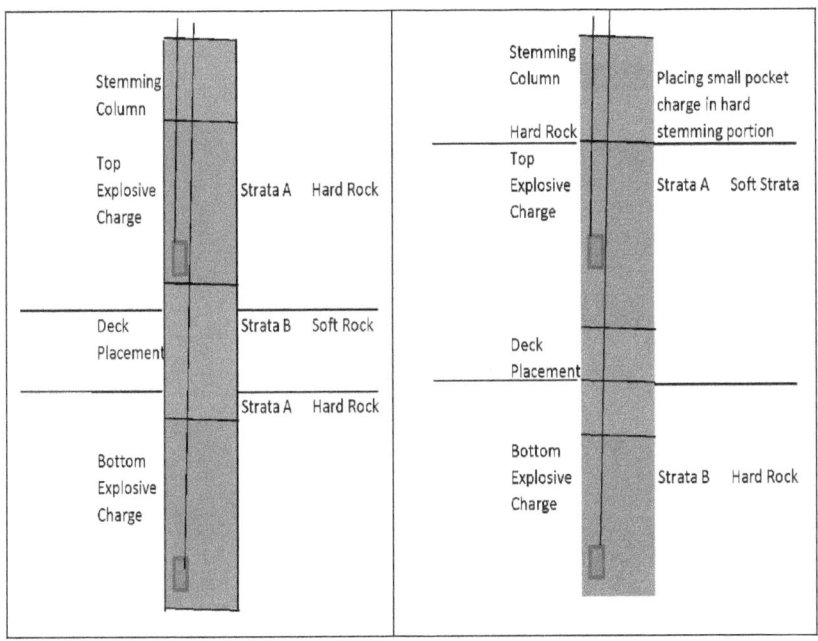

Presence of pre-formed boulders in the strata

Due to presence of preformed boulders in the given strata, this patch when blasted will generate number of boulders as these pre formed boulders will come on face post blasting. They may break down due to shock of explosive, if the holes cross through these boulders and explosive is in contact with them. Most of the time these boulders are handled by secondary blasting.

Preformed Boulders

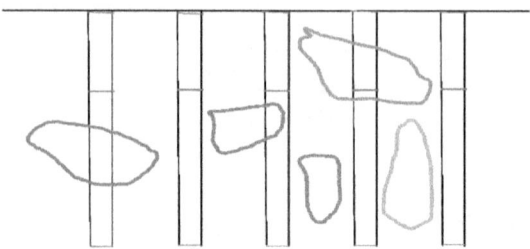

Presence of water

Water is also an important criterion for selection of explosive. ANFO is not resistant to water and this is the reason why emulsion and slurries are most wide used because of their resistance to water intrusion. If in the strata water is stagnant and not in motion, then emulsion are very stable and can with stand the condition and will blast even afters days.

But if the water is in motion in the strata, there are great chances that explosive will be washed out and blast performance will not be up to the mark. This issue is very important in the lower seam where water in the hole/strata is in motion and some time even coming out from the hole like an aquifer. In such case the washing out of explosive is common and quality of fragmentation was not up to the mark.

Reactive ground

Reactive ground is due to presence of some material that is present in strata that can react with emulsion and increase the temperature of hole, and thus can cause premature blast of detonators (if used as initiator).

Theory of rock breaking during blasting

There are number of theories that aim to explain the phenomena of rock blasting and none of the theory totally satisfies all the blast results. Yet the most commonly adopted theory deals with the work done on rocks by 2 components of explosive energy, namely Shock energy and Gas energy.

Once the explosive in a hole is detonated, there is generation of huge energy in the blast hole. This energy can be classified into two types, first is Shock Energy and second is gas energy.
Firstly shock energy is transmitted to the rock mass in the form of shock wave. The rock immediately around the charge is crushed under compression because of shock waves that are propagating outward from blast hole, but energy of these outward going shock waves is dissipated at very high rate. Soon energy of the shock wave is not sufficient to crush rock under compression (as compressive strength of rock is very high). These shock waves travel along the rock in all direction, and get reflected back when it meets a free face, discontinuity. Now as it is reflected, there is phase change by 180^0, hence the compressive shock waves are converted into tensile waves. Since tensile strength of Rock is very low as compared to compressive waves, the rock mass gets fragmented into large chunks of rocks. Also the cracks generated by shock waves are micro-fractures which exists along the entire burden

Another energy released by explosive detonation is gas energy. Detonation of explosive releases huge amount of gas and since it is in confined space the gas pressure is very high. This gas pressure causes fragmentation of large chunks of explosive generated under tensile stress into small fragments and causes movement of muck pile.

Low explosive has high gas energy but there shock energy is very low.

High shock energy is required for explosive which are used for breaking unconfined rocks but in boreholes or confined rocks, gas energy plays very vital role in blast performance.

Cross section of a hole, post blasting has been shown in figure below.

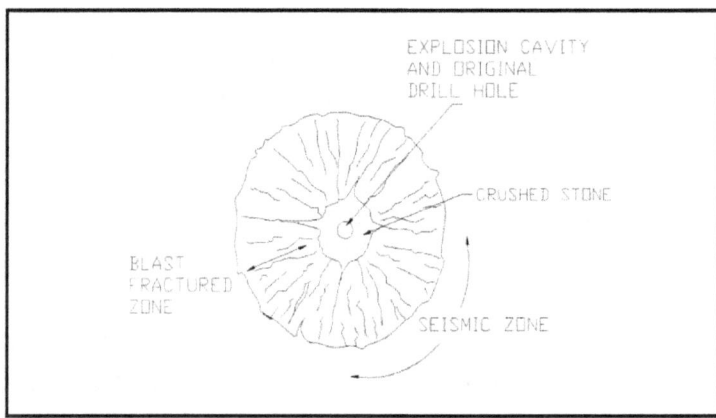

(Figure 2-1) The mechanics of blasting.

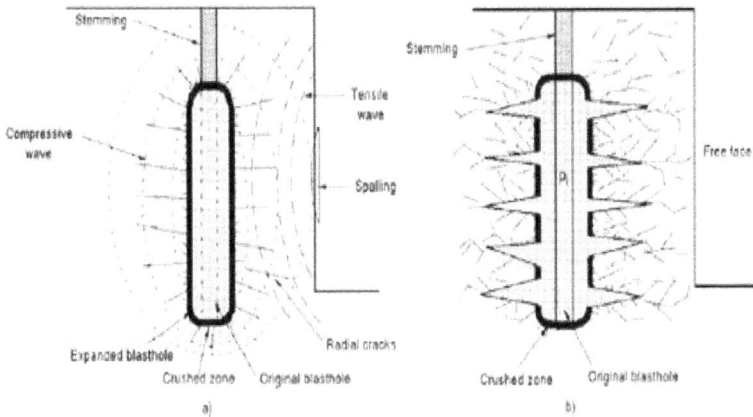

a) Shock wave propagation, b) gas pressure expansion

Blast Design Parameters

Bench Height

Bench Height is the height of the bench that is maintained for easy lifting of the material. The bench height can vary and depends on number of factors

    a> Digging height of the excavator

    b> Stratification of seams, especially in coal bearing strata.

Bench height is responsible for the spread of material of the blasted material.

Blast hole Diameter

Selection of Blasthole diameter is an important selection which to be made and accordingly drill are to be designed which requires good capital investment. Blasthole diameter is dependent on number of factors such as:

a> Type of explosive and its critical diameter: Though this is not an issue in opencast mines as the diameter is much more than critical diameter. Still the higher the diameter of explosive confinement, the better is VOD (until steady state VOD is reached)

b> Bench Height: It is common practice that higher the bench height, higher is the diameter of blasthole required.

There are few empirical formulae for determining diameter and bench heigh

$d_{min}= 10 H$

$d_{max}= 16.67 H + 50$

where d is blasthole diameter (mm), H is bench height in (m)

Burden

It is the distance that charged hole has to break from free face. This is single most important blast design criteria that controls almost every post blast result. It will decide flyrocks, air blasts, fragmentation, vibration and hence burden must be very carefully determined.

Burden changes with firing pattern. If blasting takes place such that a row of holes are blasted (line firing), then burden will be the distance between rows. But in cases other than line firing of blastholes, the burden is reduced. In case of diagonal firing, burden is really 0.7 times the burden in line firing.

Spacing

It is distance between holes that is perpendicular to burden. It can also be said that if patch is blasted in rows, then distance between the holes in same row is called spacing.

Usually there is a proper arrangement of holes in an array and this pattern is good as it is convenient to provide delay for proper movement in the blast face. On the contrary there are some cases when the blast patch is not regular and hence drilling pattern may be irregular. Still the concept of burden remains the same, which is the length of OB that has to be broken by charged hole is burden for that hole.

Fig below is self explanatory

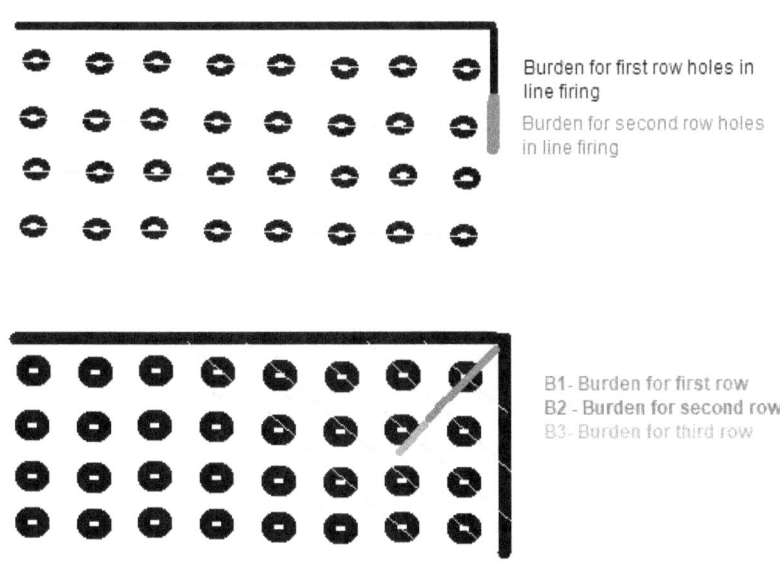

Burden for first row holes in line firing

Burden for second row holes in line firing

B1- Burden for first row
B2 - Burden for second row
B3- Burden for third row

As can be seen in case of diagonal firing (in case of square pattern) burden is a/sq root(2) i.e. nearly 0.707 times as in case of line firing (a is burden in line firing). On other hand spacing will increase and will be 1.414 times previous spacing.

If burden in blast face is reduced and charge of explosive remains same then fragmentation improves, but then cost of drilling and blasting is increased considerably. Hence cost optimization in drilling and blasting is very important. If the free face burden is very less, then one has to charge explosive in first row in very controlled manner to prevent flyrocks but that leads to generation of boulders in the front row, hence optimum burden for the first row of holes must be maintained.

If spacing is reduced, then there will be linking of holes along line and chances of gas leakage increases, thus reducing the proper throw and movement of muck pile. On the other hand if spacing is too high, then muck pile will not be uniform rather it appears as if number of hills of blasted muck is formed and portion in between two holes would not appear as blasted. In simple words the rows will not be connected and there might be case of boulder formation in face. Hence the important thing is that hole pattern should be optimized for good result that provides a good fragmentation and cost of Drilling and blasting is low.

Though there is no fixed relation between various drilling parameters but using number of field trials and with experiment it has been observed that for optimizing drilling and blasting operations following relation is used in mines.

Burden (B) = 15-25 times Diameter of hole (D)

Spacing (S) = 1.2 to 1.8 times Burden (B)

Bench Height (H) = 1.5-4 times Burden (B)

Now consider a case if D=159 mm so for sandstone strata.

B= 2.5 m to 4 m

S= 3m to 5 m

H = 4.5 m to 10 m

Sub grade drilling: Consider that a bench is to be blasted keeping the bench height of 5 m. If the hole drilled is 5 m only, then there will be some hard portion left in between holes and thus some extra depth is required in to prevent toe formation. Sub grade length changes from mine to mine and patch by patch as it is controlled by rock strata, discontinuity, bedding planes and others. The required depth of subgrade is known through working experience. For softer patches like coal seam with distinct parting, minimum subgrade of even 0.5 m is also sufficient. But for harder strata, subgrade length can vary from 10% to 20% of bench height. Thus in dragline benches 2-3 m sub grade is very normal and this reduces the chances of toe formation. There are number of factors determining the length of sub grade hole, like strata type, drilling pattern etc.

Sub grade drilling must be optimum. Too much subgrade in dragline bench can crush the coal top layer.

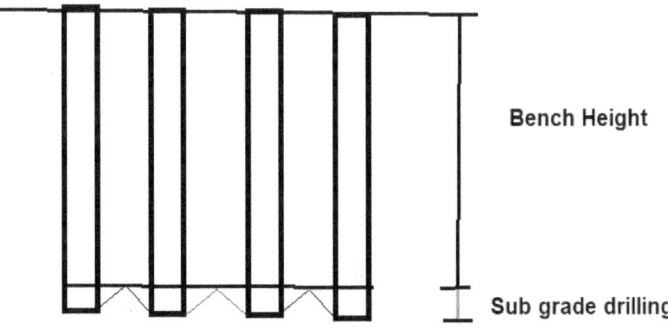

Bench Height

Sub grade drilling

Hole Depth= Bench height + Sub grade depth

Cross section of hole

Blasthole length
Blasthole length = Bench Height + Sub grade Drilling length

Stemming Column and charge column
Stemming column is the portion of blast hole that is not charged with explosive. That part is filled with stemming material like drill cuttings, sand or even water and is used to pack the explosive column. Stemming column plays a very vital role in rock fragmentation. If stemming column is more than required then there will be generation of boulders and if it less, there will be fly rocks. Hence optimization of stemming length is very important. Stemming length depends on hole depth, type of strata, drill pattern (burden and spacing) availability of dirt bands, faults, geological discontinuity etc. Stemming column should be increased in case any machinery is available near to blasting face.

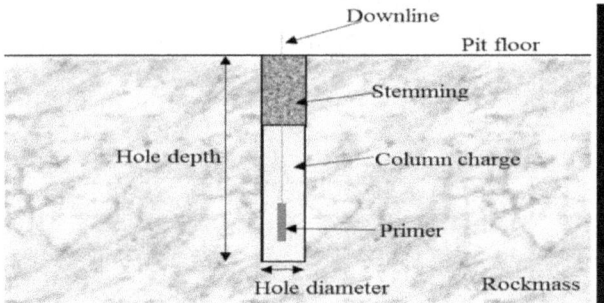

Deck Charging
Decking is a way of charging in which continuous charge is separated in two or more charge by placing a layer of OB. This is done for the following reasons.

1) Presence of loose strata in between two hard strata, like presence of dirt band etc. If the hole is normally charged, then there is possibility of leakage of energy and gas from this softer zone, hence increasing the

chances of boulder generation, toe formation and other related issues.

IN such cases if deck is placed in such manner that it lies on that dirt band, the blast performance will improve.

2) Reduce the vibration: We know that vibration depends on the maximum amount of charge per delay blasted in a round. By placement of deck the maximum charge per delay is significantly reduced if top and bottom charges are initiated in different intervals. Even if top and bottom charge is blasted at same instant yet total explosive is also reduced, hence reducing maximum charge per delay.

3) Decking helps in proper utilization of explosive. For example by decking in DL benches one can easily save nearly 200 kg of emulsion explosive (for diameter of 269mm and depth 30m) per blast hole

In case of deep hole, the amount of explosive required is much less than what is provided.

Volume Calculation

Volume of the muck generated by blasting 1 hole= Burden * Spacing * Bench Height

Total Volume of blasted muck = L*W*BH

Where,

L is Length of Blast Patch (Hole to Hole + Spacing) in m

W is width of Blast patch (Hole to hole + Burden) in m

BH is bench Height

Powder Factor Calculation

Powder factor is defined as the Volume of OB or Tonnage of Coal blasted by 1 kg of explosive.

So PF= Volume of Rock blasted (or Tonnage of Coal)/ Amount of Explosive used

Even achievement of Powder factor is used as criteria for quality of explosives, and relative deductions are made if powder factor is not achieved.

For determining Tonnage, Simply multiply Relative density of Coal/Ore with Volume blasted (in m^3).

Few of the blasting techniques that can be used for improved benching and prevention of backbreak and used in some mines are:

a: Presplitting : In this technique a fine crack in rock mass is created in the high wall so that there is no back break generated due to blasting in massive rock mass and high walls are stable. This also reduces vibration significantly as this crack will stop vibration because of discontinuity.

In this technique holes of depth equal to bench height are drilled closely and spacing between holes is given as

S=10* d

Where d is diameter of hole,

These holes are very lightly charged and along the length of holes, so that after excavation the profile of highwall is smooth and stable. One important point in presplitting is Half cast factor and percentage of half cast.

In massive rock, half cast holes or half of boreholes are visible in highwall after removal of muckpile. Effectiveness of presplitting can be measured using half cast percentage which is simply the % of half cast visible to total presplit holes. Though it is very vague concept because in geologically disturbed strata or soft strata, half cast hole may not be visible, despite good effort.

Powder load is the charge provided in the presplit holes per m length of the hole.

For charging of presplit holes slurry explosive in form of cartridge of small diameter can be used and connecting along entire length using Detonating cord. These presplit holes are blasted before blasting of the production holes.

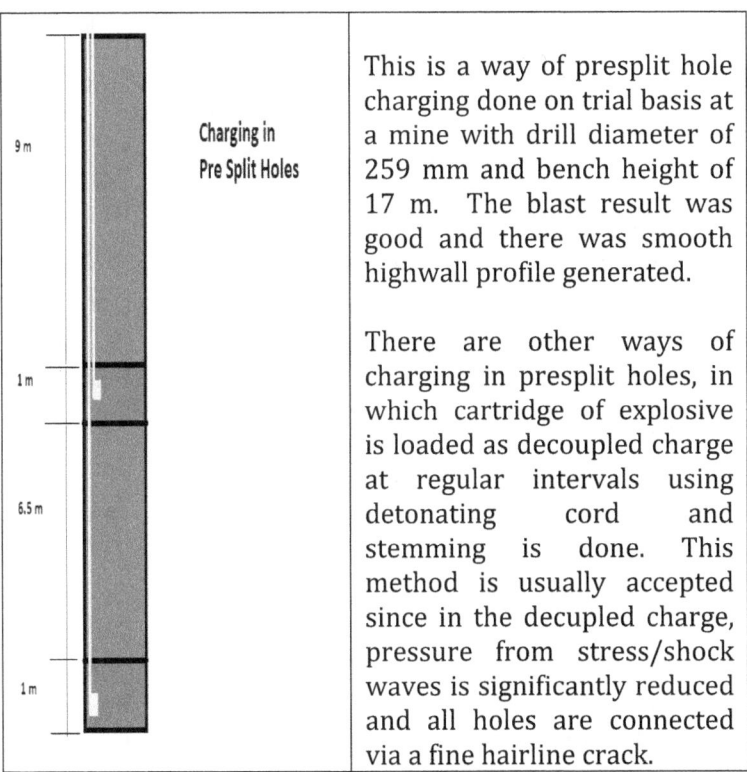

Charging in Pre Split Holes (9 m, 1 m, 6.5 m, 1 m)	This is a way of presplit hole charging done on trial basis at a mine with drill diameter of 259 mm and bench height of 17 m. The blast result was good and there was smooth highwall profile generated. There are other ways of charging in presplit holes, in which cartridge of explosive is loaded as decoupled charge at regular intervals using detonating cord and stemming is done. This method is usually accepted since in the decupled charge, pressure from stress/shock waves is significantly reduced and all holes are connected via a fine hairline crack.

b: Line drilling: This technique is similar to pre splitting technique for smooth highwall except for the holes are not charged and spacing between holes is further reduced.

c: Muffle Blasting: Muffling or covering of Blasting patch with steel mat and then loading the bags filled with sand on the blast hole. The primary objective of muffle blasting is to minimize fly rocks. This technique is used when blasting is done in vicinity of dense human population area.

The primary consideration for drill hole is that the length of hole should be uniform and slightly more than desired bench height. There are issues which we face some time especially in shovel benches where shovel moves up. This is because of small portion which is not properly blasted and is not cut by shovel. In that case we need to cut small hard portion using dozer, so that bench height is maintained.

Bench Stiffness Ratio

Bench stiffness ratio is defined as the ratio of bench height to burden ratio. IF BSR is below 2, then the blast patch is tougher to blast. In this case, one has to increase the sub grade drilling for the patch.

When BSR is 2-5 times burden, then the stiffness is reduced and blast has a better throw and good muck pile distribution. For BSR higher than 5, there will be good casting of material. Cast blasting should preferably have more BSR.

Hole Inclination:

It is also an important parameter in blast design. Inclined holes are typically very helpful in case where there is high toe burden.

Drilling

There are 2 forms of rock breakage

a. Rock penetration
b. Rock fragmentation

Drilling is one method of rock penetration and it can be divided into 2 common modes based on the application of mechanical energy applied i.e. Percussive drilling and Rotary drilling.

In percussive drilling the penetration in rock is done by the effect of successive impact applied through chisel shaped bits. Hence in percussive drilling 2 dominant mechanism are crushing and chipping.

In case of rotary drill the constant pressure in applied on the drill bit, and the bit is rotating, which causes crushing, cutting or abrading of rock as per the rock strata.

In surface mining Roller bit rotary drilling is mostly favoured drilling mechanism.

Great part of modern drill is that there is a pressure monitoring system at various depths. So during drilling if there is any strata change then the feed pressure will always change and one can find coal or OB depth variation. During drilling air pressure is used for removal of drill cutting. Drilling becomes extremely difficult in loose or fault zone strata and many times rod gets stuck. Uniform strata, even if hard, rate of drilling is good, but if there is some loose strata, strata variation, presence of geological discontinuity, rate of drilling is reduced and there is more consumption of drill bits (higher abrasion of bits).

Line drilling is done at the high wall side. This was done to prevent back breaks and to arrest ground vibrations.

There are many instances where the formation of boulder occurs because of these geographical conditions.

Accessories used in Blasting

1. Boosters-Cast Booster and Emulsion Booster
2. Detonating Cord
3. Detonating Relays (Cord Relay)
4. MS Connectors
5. NONEL- DTH and TLD's
6. Electronic Detonator
7. Electric detonator
8. Safety Fuse
9. Cartridge Slurry
10. P-3 explosive for Boulder breaking
11. Exploder and cable

Initiating system

Initiating systems are arrangement done for conducting a blast with required delay sequencing. Primary purpose of initiating system is to provide necessary shock to the prime charge/booster for blasting of bulk explosive. In the first case the initiating system will provide fire/current so that the signal is transferred in the lines where all blast holes are well connected. The fire/signal is sent down the hole using initiating system component and this provides energy for initiation of primer charge.

Components of Initiating Systems
1) Source of initiation
2) Transmission of signal to each blast hole
3) In the hole initiation

Classification of initiating system for blasting can be done according to these components and is very varied.

Based on source of energy for initiation of blast, Initiating system can be classified as

Initiation by electric detonator- Detonating Fuse, and NONEL (shock Tubes can be initiated using plain electric detonator;

Initiation by plain detonator and Safety Fuse- Detonating Fuse, and NONEL (shock Tubes can be initiated using plain electric detonator;

Initiation of Electronic Detonator initiating system using their designed blaster;

Based on transmission of energy for blast each blast holes and in the hole delays, initiating system can be classified as

1) Detonating Fuse based and delay elements are added by Surface relays or MS connectors

2) Shock Tubes- Surface Trunk lines (STLs) are used and at the end of each shock tube is connected a delay detonator. STLs are connected with the Down the hole (DTHs) shock tubes that are attached using slot provided in STLs. Also the DTHs have their own delay detonators. This helps in providing delay to each blast holes.

3) Electronic detonators have harness surface wire that is connected with each blast holes and does not require any external shock. The bottom of leg wire has its own electronic detonator that has flexibility of adding any desired delay timing.

Detonating Cord/Detonating Fuse (DF)

Detonating cord is a plastic cord inside of which is made from PETN (Penta erithrotol tetra nitramine). Amount of PETN in DF varies from 5 to 10g per meter length of DF. DF is filled with PETN and reinforced with yarn for taking loads. The cast booster is directly tied to DF and primed in the hole. At the top of hole each DF is connected using DF as trunkline and blasting is done by initiating DF from an electric detonator. VOD of DF is nearly 6500-7500 m/s and provides sufficient energy to initiate cap sensitive explosive. There are some issues with the use of DF.

As it detonates, it produces a huge noise when blasting is done. This creates a psychological pressure on the nearby living people. Other initiators like NONEL or E-dets do not create any noise during blasting.

Another major issue is that DF burns the emulsion explosive or destroys the gas bubbles of some portion of the explosive column that is in direct contact with DF inside the blasthole, thus reducing available energy for blasting operations. One can easily reduce 2-5 kg of bulk explosive per hole of average 5m depth, just by switching initiation system from Detonating Fuse to NONEL.

Since DF detonates as it blasts, hence there is case of loosening of stemming of blasthole. In watery holes, blasting using DF leads to more number of blowouts, and using NONEL reduces this chances of blowout as stemming is intact and the hole is bottom initiated.

Another advantage of NONEL or Electronic detonator is that they provide bottom initiation. With the introduction of bottom initiation system issue of toe formation which was serious problem at high depth benches like dragline, when blasting was done using DF.

Detonators

Detonators are used for initiating high explosives and contain small amounts of a sensitive primary explosive. Although they are manufactured to absorb a reasonable amount of shock during handling and transportation, they should be handled very carefully. In general detonators consist of an ignition charge, intermediate charge, and a base charge. Each charge in the series is selected and used for transition from heat to shock.

a. Non Electrical Detonators

These are used to initiate other explosives, detonating cord and shock tube. These are composed of Lead azide, Lead Styphnate and Aluminium powder (commonly termed as ASA) as primary charge and PETN and RDX as base charge in the detonating cap. Non-electrical detonators or fuse caps are thin metal or paper cylindrical shells, open on one end for the insertion of safety fuse or NONEL shock tubes which contain various types of primary and secondary explosives. They are sensitive to heat, shock and crushing and are designed to be initiated with safety fuse or detonating cord or shock tubes. They are normally rated as #8 strength detonators. Detonators of this type can be instantaneous or with delay detonators. Shell of the detonator is made of copper or Aluminum.

With aim of reducing Lead Azide as primary charge because of its contaminating properties to environment, other chemicals are also used such as NHN (nickel hydrazine nitrate) or BNCP (Bis (5 Nitro tetrazoleto) tetraamine cobalt perchlorate or DDNP (Diazo Dinitro Phenol).

b. Electrical Detonators

These are used to initiate other explosives, detonating cord and shock tube. These may consist of RDX/PETN, ASA. Electrical detonators are similar to non-electrical detonators except they are initiated by the application of electrical current through electrical wires. The current causes a bridge wire or match elements to heat/function thereby, causing the ignition charge to explode which in turn, causes a chain reaction to cause the base charge to be initiated. The wires are secured into the detonator by a closure plug, crimped into the shell, which seals the explosive from moisture. In addition to sensitivity from heat, shock and crushing, these products are subject to extraneous electricity due to the presence of electrical wire.

Detonating Relays:

These are non electric surface delay detonators to provide delay between holes that are connected
along with DF. They are provided with grooves. DF is cut and desired relay is provided by tying both ends of DF with the relays. One can fire desired pattern using relays and DF like line firing, diagonal firing, V shaped firing etc. Relays provide various delays like 25ms, 50 ms, 100ms etc.

The main disadvantage of using relays is that their delay timing is not accurate. Hence there are great chances of overlapping of blastholes thus leading to all sorts of blasting issues like back brake, fly rocks, vibration, generation of boulders and everything. Though we can control overlapping by providing sufficiently long delay between rows and blast performance will be satisfactory, yet for long patches timing of blast will increase that will lead to increased time of vibrations and that may create unnecessary friction with people living nearby area.

MS Connectors

MS connectors are bidirectional detonators used for providing delay when DF is used as initiating system. They consist of shock tube, which are crimped with detonator at each end. They are colour coded and provide better delay timing, as compared to relay detonators as the scatter is for less. Balancing of delays in dragline using MS connector is important, so that delay timing for hole shall be same.

NONEL

NONEL initiation system stands for Non Electric detonator initiation system. NONEL consists of a detonator which is initiated using shock tube. Inside the shock tube is a very thin inner lining of PETN explosive. As this shock tube is initiated, the explosive inside the tube starts to deflagrate at VOD of nearly 1800 m/s. Since there is no detonation of explosive, the plastic tube remains intact, and there is no noise on surface (as in case of DF). NONEL that we use consist of 2 components namely DTH detonators and TLD detonators. DTH detonators are used for priming for the bulk explosive (Booster or cartridge) and are put inside the blasthole before charging of Bulk Explosive.

View of blasting using NONEL is very soothing to watch. The Trunkline delay detonators initiate every Down the hole detonators and each hole is blasted at different time (at a gap of few ms).

Since the DTH detonators have high delay (250-400 ms) and TLD have lower delays (usually used 17-65 ms), hence after maximum blasting of TLD detonators only then there will be ground movement.

There can be one case that DF is used for surface connection and priming is done using DTH detonator. In this delay is provided using relay or delays. Firing pattern can be line firing or diagonal or any other as per requirement. In this case there is no precaution required as blasting is same as using TLD.

If there is case when in some patch some holes are primed using DF and some using DTH NONEL, in that case since there is delay in DF, and 250ms delay in DTH, so front rows should be primed using DF and back rows to be primed using DTH. There will be some boulder formation due to change in delay timings but overall blast will be easily mucked. On the other hand if back rows are primed using DF, then back rows will blast first and front row will blast later. This will lead to generation of boulders in entire patch as face burden will be too big and explosive from DTH blastholes will leak out hence producing more of boulders.

One very serious issue with NONEL initiation system is that if NONEL is stretched (during priming or mishandling), there is formation of gap in PETN layer, because of which propagation of shock inside tube gets disrupted and detonator may fail to fire, thus leaving live detonator and cast booster and explosive intact in the same blasthole (Misfire). There can be of blasting of this live detonator, while mucking (person were not aware that they were lifting material and there was a case if misfire) and damage to shovel and injury to operator because of this reason.

Another issue with any detonator initiation is that should not be charged in hot holes. Since the detonator has a primary charge, which is very sensitive, that can detonate due to high temperature, hence in that cases use of DF is best alternative. Also use of NONEL doesn't provide desired result in cracked strata patches because of multiple loading and decking of the hole, because of which there are many chances of stretching of NONEL.

There are few tests that are done to check the quality of NONEL shock tubes.

   a) Tensile strength and elongation Test: In this test, the quality of tube is tested by an instrument called tensiometer, in which NONEL is stretched until it snaps.

The Tensile strength is the expressed in N (Newton) and longitudinal elongation at break is in %.

b) Thermal Shrinkage test: This test is done to estimate extend to which shock tube can be shrinked when exposed to high temperature. This is property of polymers that when exposed to heat they tend to shrink. Testing is done at nearly 80^0 C.

c) Oil Ingress test: Since the NONEL is completely immersed in bulk explosive which has oil as its continuous phase, Shock Tubes are tested if there is any misfire due to oil ingression. For this a bath of fuel oil is taken and shock tube is immersed for 50 hours and then tested if misfire occurs.

d) Burst Test: In this test the shock tube is heated for some time at specified temperature and then initiated. Number of burst per m of shock tube is recorded and it provides estimate for radial strength of the shock tube.

e) Velocity of Detonation:

Beside these tests done for quality of Shock tubes, test of detonators are also done

a) Delay scattering Test
b) Impact Test
c) Thermal Sensitivity Test
d) Impact Sensitivity test

Electronic Detonator

This is the most advanced initiation technique of blasting rock. Use of Electronic detonation provides all advantages of NONEL with extra benefit of lower delay scattering and flexibility of delay timings thus reducing and controlling vibrations, back breaks, fly rocks etc.

E-det have programmable digital detonator which contains microchip energy storage capacitor, safety structures and conventional explosive components. The microchip circuitry includes an oscillator for timing, memory for retaining its programmed delay and communication functions to receive and deliver digital messages to and from the control equipment.

The electronic detonator system consists of following key components

        a.   Digital detonator

        b.   Connecting wires

        c.   Logger and Blaster.

Typical characteristics of an electronic detonator are.

- The detonator initially has no initiation energy of its own
- The detonator cannot be made to detonate without a unique activation code
- The detonator receives its initiation energy and activation code from the blasting machine
- The detonator is equipped with over voltage protection. Low excess loads are dissipated via internal safety circuits and higher voltages limited by means of a spark plug
- The initiation systems operate with low voltages, which is a great advantage considering the risk of current leakage.

Delay time allocation to the detonators is carried out by uniquely coded signals to eliminate any possibility of error. The detonator responds to the code only from the blasting machine and thus eliminates any risk of initiation from external energy sources. The blasting machine also performs an operation status control, which is done automatically by the machine.

Each detonator can be programmed for delay interval anywhere from 1ms to 8000 ms. The blaster communicates to each detonator in turn via the logger. Each blaster can handle 8 loggers and each logger can take 200 detonators, giving a system capability of 1600 detonators per blast. At any time connected detonators can be tested and there is a full two way communication between detonators and control equipment.

Advantages of Electronic detonators over other initiation system:
Bottom Initiation
    Less chances of Toe formation
    Reduced chances of fly rocks

Precise delay timing without scattering effect
    Proper fragmentation
    Reduced vibrations
    Reduced flyrocks
    Less back breaks and side breaks

Stemming retention time is higher, so heaving effect due to gas energy is higher.
    Higher utilization of explosive energy
Safe from extrinsic charges

Basic Detonator Construction

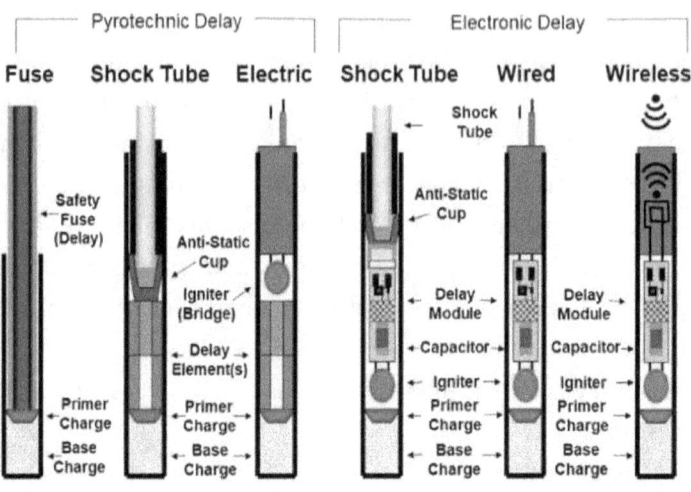

Electric Detonator and Safety fuse

Any blasting patch wherein initiating system used is DF or NONEL, the surface connection is done and line of DF or NONEL is extended, and a detonator (electric detonator or Safety fuse) is attached to fire the blast patch.

Safety Fuse: This is used to initiate non-electrical detonators and is composed of black powder as core material with plastic tube and layers of fabric to provide strength and flexibility of rope. It has fixed burning rate and this initiates non electric detonator, which is connected with DF or NONEL (TLD). Safety fuse burns at a rate of approximately 35-45 ms/ft. Safety fuse is constructed with various types of natural and manmade fibers and plastic.

Electric detonators: In case of Electric detonator, there is a wire through which electricity is passed and as the wire is heated,

Primer

Primer is the unit charge (Cap Sensitive), along with initiator, that is needed to provide energy (initial shock) for the initiation of the bulk explosive (Column charge which is non cap sensitive). Without the primer of sufficient strength the initiation of bulk explosive will not take place and performance of blast will not be satisfactory. Primer charge are usually cap sensitive explosive that can be easily initiated using No 8 cap detonator or using detonating fuse.

There is a difference between primer and booster. A primer is when booster is attached with initiator. Else boosters are placed in the charge column to provide the additional energy into the hard strata.

There are many criteria for determining the type and quantity of primer used in the explosive charge. The selection criteria are composition of primer and shape of primer. Composition determines the detonation pressure that impacts initiation of main charge. Also very small diameter primer doesn't interact efficiently with the column charge. Low energy primers used will cause burning reaction in column charge and detonation will not take place. A number of instances have been seen that bad quality boosters have caused the failure of blast.

Role of VOD of primer on VOD of explosives has already been discussed.

CARTRIDGE

It is composed of Slurry explosive.

These are of 2 types

a: Prime Charge: those which are cap sensitive. In case of prime charge, paint grade aluminum is used as against coarse grade used in Base charge. This increases sensitiveness.

b: Base Charge: these are non cap sensitive, hence cannot be directly initiated using 8 strength detonators.

Before the use of Bulk explosive these cartridge were the explosives used. Handling of these cartridges manually, was labor intensive process. Even in small mines, where requirement of explosive is less, cartridges are used as bulk explosive.

Cast Boosters

Cast boosters are made of Pentolite. Pentolite is a mixture of PETN and TNT. These are easily detonated using Detonating fuse or detonators. Their VOD is more than Emulsion explosives and nearly 7000 m/s. Cast boosters provide sufficient energy for initiation of bulk explosive. According to some research it has been proved that optimum cast booster required for initiation of a charge column is nearly 0.2% of total explosive in a charge column. So if 50 kg of emulsion is used then we will require nearly 100 g of Cast booster.

Standard practice for using cast booster is that 400 g of cast booster (only 1 CB of 400 g) is used for initiation of the charge column despite what be the amount of explosive in the given charge column. Cylindrical cast boosters are most common but there are different size and shapes of cast boosters available in market. Researchers have found that conical and spherical cast boosters provide better initiation shock energy hence providing improved results.

VOD measurement of Cast boosters is done using same method as that of cartridge or bulk explosives using the D-Autrich Method.

Emulsion booster are made of emulsion explosive that are cap sensitive and can be detonated using number 8 detonator.

Emulsion booster is a cheaper substitute of the pentolite based cast booster. The only issue with using the emulsion based booster is that the VOD is very less and, hence for the explosive overdrive zone will be higher. This will waste explosive energy of SME Bulk, which could have been better utilized.

Other Accessories

Blasting Cable

Blasting cable is required for connection of electric detonator to the initiating system such as NONEL or Detonating Cord. Electric detonator (zero delay) is connected to the zero number blast hole (First hole that is to be blasted) and electric cable of sufficient length is laid so that the blasting mate is safely inside the blast shelter (nearly 200m from blast patch). Resistance of wire should be low and continuity must be there. They must be flexible and insulation must be with adequate strength so that can be used reliably in robust condition and water (as per the mines condition).

The required electric charge is provided by exploder.

Exploder

Exploder are the machines that provide the required electric power to fire the series of detonators or the single electric detonator that is connected to the initiating system.

Measuring Tape

This is required for measuring of depth of blast hole, measuring the length, width and bench height of blasting patch. Tape must be of adequate strength and flexible sufficient strength so it doesn't break with slight load. At many places digital depth meters are used, which can save much time and energy.

Measuring Beaker

A measuring beaker of 1 L capacity is available for measuring the density of bulk explosive and rate of gassing reaction in the field during bulk charging activity. The density of explosive in cup is measured at different interval to measure the rate of gassing. Initial cup density is almost fixed but the desired final density is varied as per depth of the blast hole.

Spring Balance:

This is required for measuring the density of bulk explosive during charging. These must be properly calibrated.

Spades and Shovels for stemming

Tamping Rods

Blasting Knifes or pliers

Warning System during clearance

Charging Operations

1) Accessories are brought from magazine to charging face.
2) Booster and Initiating system such as NONEL, Detonating Fuse or Electronic detonator placed near

each blasthole, and each hole is primed by inserting the initiating detonator or Fuse in the booster and lowering the primed booster inside the hole. If the position of primer is in the top of explosive charge column, this is called Top Priming and if the position of primer is in the bottom, then it is called bottom priming. Bottom priming usually provide better result than top priming, as initiation of bulk in bottom priming is from lower portion, thus providing more time for the gases to move toe burden, whereas in top priming the gas energy generated is easily leaked out from fractures and cracks, thus less time is available for gases for movement of toe burden. But in case of hot holes, priming should not be done as that may initiate detonator or fuse, and priming is done only after charging of bulk explosive (as explosive doesn't detonate because of temperature in range of $100\text{-}200^0$ C. Top priming is also practiced when top portion of bench is very hard and lower portion is softer, thus energy is available for breaking top hard portion. Cast boosters are always preferable (except in hot strata) for initiation of bulk explosive as it provides required energy that makes bulk explosive reach its steady State VOD in shorter charge length thus improving blasting result. Emulsion booster doesn't provide that level of energy hence steady state VOD of explosive is not reached instantaneously, thus reducing blasting performance (Details are discussed later). People engaged in charging and blasting operation must have proper PPE kit, i.e. Helmet, anti-static safety shoes, clothes should be of cotton which should not generate static charge, gloves etc.

In case of bottom priming, a precaution must be taken that the position of primer must not be at the bottom, rather it should be located slightly higher. Reason behind this is, in case of watery holes, there are chances that bottom of hole has mud, and primer if placed in mud, doesn't get properly connected with explosive, and may not initiate the charge itself. Another benefit is in case of subgrade, the chances of leakage of energy are reduced if the primer is above transition zone.

Also in deep holes, one can go for multiple priming. In this case priming of explosive charge is done at bottom, and additional boosters are placed at some interval. This help retain the VOD of explosive charge column, though this can be additional overhead cost.

In case of deck charge, separated charges, multiple priming is compulsory for initiation of bulk explosive.

3) After priming is done charging operation of bulk is to be started. Firstly BMD must be checked if there is some visible issue/problem. If there is any issue of overheating, over pressure or any leakages operation should be instantly stalled, and charging should only start when this issue/ problem is resolved. All necessary valves such as for water, gassing agent and emulsion must be opened. Earth chain must be grounded and wooden blocks must be placed to prevent movement of truck in any slope. Parking brake must also be applied. Any features related with safety should not be compromised as that may lead to very serious accidents. BMD must be parked in such a manner that it doesn't run over any blasthole. Drill patch must be provided in manner that there is ample space for parking of BMD. Usually length of delivery hose is 40m

but sometimes, this length could be less in BMD. Length and diameter of delivery hose is also very important aspect of safety. Diameter of delivery hose is 2 inches or 1.5 inches. If the length of delivery hose is very long, that will create additional pressure in the positive displacement screw pumps and will frequently trip the BMD system (because of the cut off pressure) and too small hose length will create problem in charging of face as BMD might have to move inside the blast patch.

Sometime especially in rain, drilled patches are not even reachable by BMD so longer delivery hose is required from mine authority. Small diameter delivery hose also increases the pumping pressure and because of that frequency of BMD breakdown increases considerably because of small diameter delivery hose.

BMD delivery hose is designed in such a way that it is anti- static in nature i.e. if any static charge is accumulated during charging operation in BMD, then delivery hose will discharge any accumulated charge into the ground. This is done by providing steel wire threads in the delivery hose. Also the diameter of hose is less than critical diameter of bulk explosive, so in case if explosive is initiated during charging process then also explosive in hose will not detonate, though this has never happened.

4) After properly parking the BMD vehicle near blasting patch and applying parking brake, and dropping earth chains, valves are opened for charging. These valves will control flow of gassing Agent and water. Delivery hose should be unfolded from hose drum on surface before charging. Depending on the requirement of charging face, amount of emulsion per hole is charged. Each drill

hole is measured before charging and explosive column is measured. It is very important to measure each and every hole so that there is no issue of overcharging or undercharging. Loading Sheet is filled simultaneously and quantity of explosive charged per hole is noted, initial and final density is recorded and length and width of patch is measured. With this information, we can determine the in-situ powder factor achieved for each blast.

5) GA is varied to control rate of gassing. Water flow rate is increased if pressure in system increases, as water acts as lubricant, when explosive passes through delivery hose. Emulsion pump is started to pump emulsion from bin into flow line and both the component, namely GA and emulsion are mixed together using a static mixer (in case of doped BMD, AN prill are also mixed at static mixer only). This mixture is pumped to product pump from where emulsion explosive in pumped into hose and finally into the blast hole. Water is injected just at the mouth of delivery hose in such a manner that it forms a film of water around the explosive, thus reducing friction that will be generated because of movement of emulsion in hose pipe. Flow of emulsion should be uniform during charging operation as that will produce uniform mixing of GA. Flow rate of emulsion pump and product pump should be nearly same, as to prevent any overflow of emulsion from bins. Delivery hose must be lowered inside the watery hole as emulsion will not come in full contact with primer explosive if not inserted. As per DGMS guideline, delivery hose must be inserted in all holes (be it dry or watery) to prevent shock on primer

as well as on bulk explosive. Rate of charging can be changed as per requirement by increasing hydraulic pressure using throttle valve.

6) Since gassing of product takes some time (depending on amount of GA mixed) after charging of blast hole, stemming should be done nearly after 15-20 minutes. Column of explosive rises as gassing reaction takes place in explosive. Samples of explosive must be taken to record rate of gassing in explosive. Sample of explosive are taken in container of 1 L capacity and spring balance is available to check initial and final density of explosive. Density of product is measured at interval of 3-5 minutes, and when the desired density is achieved (depends on depth of hole and strength of strata), stemming is done. Final stemming height is predetermined and holes must not be overcharged as that may lead to fly rocks. Undercharging also leads to generation of boulders. Amount of charge per hole must be closely monitored and is known using trial and error method, so experienced blasting Overman and other personnel have important place in blasting operations. Length of charge column is very important must be checked. Some times BMD works such that though the counter shows some amount of explosive and column increases more than required, there can be case of overcharging. To prevent this from happening, BMD vehicle must be regularly calibrated. Working of counter (that displays amount of explosive that goes into blast hole) works on the principle that as the rotor moves, a proximity sensor counts the number of rotation of rotor and put a multiplicative factor to determine amount of explosive charged. This is very

crude method and will vary if there is change in viscosity of emulsion, amount of explosive in bin, rotation rate of emulsion pump and product pump. This issue of calibration can be sorted out if there is provision of weighing the delivered amount (measuring mass flow rate), thus delivery counter will give exact reading of quantity delivered in hole.

7) After charging of bulk in charge holes, BMD must move out of the blast face. Since BMD has battery, engine and moving parts these can cause initiation of primer. No charging operation should be done during thundering, lightening or even during rain. Accessories must not be carried in BMDs.

8) Charging in cracked and hot strata is discussed later in the book. When desired stemming column is achieved, stemming of the blast hole is done, using the drill cuttings or sand if available. The blast holes are connected using STL (in case of NONEL), Detonating Fuse or connecting wire (in case of Electronic detonator) and desired delay is provided. In case of NONEL, STL or TLD have their prefixed delay timing and each holes blast separately, but in case of DF, blasting each hole separately will be very hectic process as each hole has to be connected by relays. Electronic detonator is the latest technology for providing delay timing in the blast holes.

9) Clearance for the blasting is taken before blasting the charged patch. All people in the nearby locality are removed and siren is blown before blasting. Only when the shot firer gets instruction from blasting manager, who looks after clearance from all areas surrounding mines, shot is fired. In case of DF or NONEL as initiating

patch, there is requirement of electric detonator which is connected at some distance from the blast patch and then copper wire of nearly 100 m is attached, from where, exploder is connected to the electric detonator and shot is fired. As per instruction from DGMS, for blasting no person must be present in 500m zone.

Delay Sequencing

What is Delay sequencing

When blasting number of holes, it is common to provide delay in blasting such that all the blast holes do not go off at once and blast in sequence such that holes in front rows blast first and there is movement of burden thus creating new free face to the hole in the back rows. This is called delay sequencing and is very important. Delays between the holes are provided by initiating systems or by surface connection. Various reasons for proper delay sequencing are mentioned as under:

Need for providing delays

1) Proper fragmentation: Progressive movement of free face for back rows thus relieving increased burden and hence improves fragmentation. If the back row holes fire before the front row holes by some chance, there will be generation of boulders in the patch as the burden for these holes will be very high. Also explosive charged in front rows will leak out from holes due to generated cracks, thus reducing explosive energy available for blasting, further increasing chances of

boulder formation. This condition can come in case the scatter of delays in the detonator is high.

In case of diagonal and V or extended V firing pattern, the fragmentation of blasted material is improved as the effective burden is reduced by nearly 30%.

2) Lower Vibration: It has been observed that with increasing the burden for a blast hole, the vibration level produced is higher. With the proper delay interval, and uniform movement of burden, so that the progressive burden is uniform for holes in back rows, the shock energy of explosive are not wasted in the form of vibration. Also by changing the firing pattern, one can easily change the direction of propagation of vibration waves, thus preventing the desired structure. It can be very easily implemented using V or extended V firing pattern, in which a charge patch is divided into two patches due to hole segregation and effective vibration in each direction can be reduced.

3) Reduced fly rocks: Improper delay or overlapping of delay may increase chances of fly rock. This can happen when hole in front row has not been blasted properly and back row has initiated. Since the burden is too high, the easier path for gases to escape would be from the stemming column and throw of material would be upward. In such cases fragmentation is not satisfactory.

4) Increased stemming retention time: Higher is the stemming retention time, higher is the time available for the gas energy to cause heaving of the material. With the help of bottom initiation, the stemming retention time is significantly increase as compared with top initiation. As stemming retention time is increased, the chances of toe formation reduces.

5) Lower side breaks and back breaks: Due to improper delay timing or overlapping of delay of blast holes, the issue of back break and side breaks is significantly increased.

In general the delay should be provided in such a way that intra row hole delay should be lower and inter row (change in row) delay should be higher.

Also according to some study, it has been found that for medium hard sandstone, the burden movement time is nearly 8 ms/m of burden. So delay for row change in outsourcing patch must not be less than 32 ms (Burden is 3-4 m) and in hole delays can be 17ms or 25ms. Commonly for 2nd and 3rd row, jumping of delay is done using 42ms or 67ms. But if the number of rows in a blast patch is higher, the jumping delay for back rows must be increased subsequently up to 100ms. Any jumping delay higher than 100ms should not be used in small depth holes (6-7m).

Adequate delay used for jumping ensures that holes in the front row has been blasted, and burden has moved, thus chances of overlapping is sufficiently reduced even if there is some scatter in the initiation system. If the delay is increased the chances of fly rock generated from back rows decreases significantly as there is free face generated due to movement to front rows but if delay is increased or not optimised, then there will be generation of boulders.

Scattering of delay in holes and its impact:

Scattering of delay for any detonator indicates the variation in the time at which detonator actually blast w.r.t. the delay it is designed for.

For example if a detonator is labelled 17ms, but there will be some deviation in time blast for this detonator. This is primarily due to variation in delay element. It is very common in non electronic initiation systems that there is scatter of delay during blasting from that mentioned by manufacturer. For example if there is +/- 2 ms-5ms scatter in NONEL shock tubes. In such cases the best practice to adopt, is by increasing the inter row delay timing. This will ensure that front row will be blasted first and back rows will move only after front row of blast holes. But too much increase in inter row delay also has poor effect as there will be chances of leakage of explosive from blast column and also cracks will reduce the heaving effect due to gas energy and may lead to generation of boulders in the blast patch.

Firing Pattern

1) Straight Line firing pattern

   In this firing pattern, holes in the first rows are blasted and holes in the back rows will be blasted. In this pattern scattered drill hole pattern is suitable.

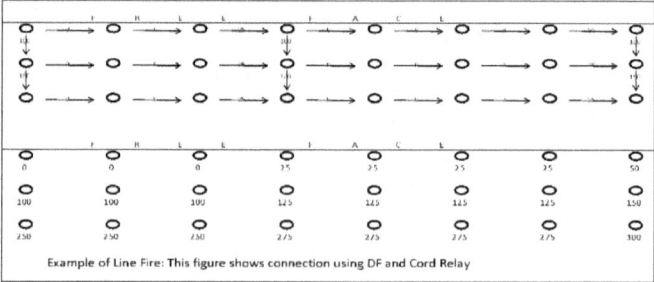

Example of Line Fire: This figure shows connection using DF and Cord Relay

2) Diagonal And Partial Diagonal Firing pattern:

In this pattern, the holes will blast diagonally. In this firing effective burden is reduced by nearly 30% as can be seen in the figure.

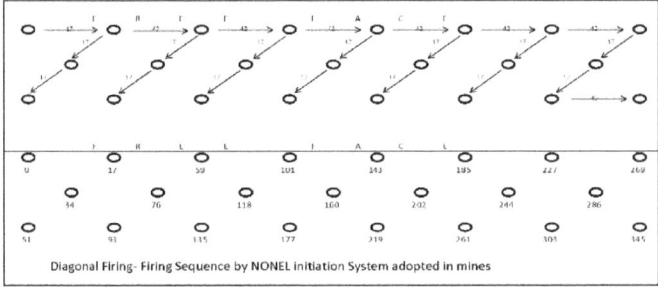

Diagonal Firing- Firing Sequence by NONEL initiation System adopted in mines

This firing pattern is suitable for blasting patches having 2 free faces.

In this firing pattern, rectangular drill pattern of blast face is preferred.

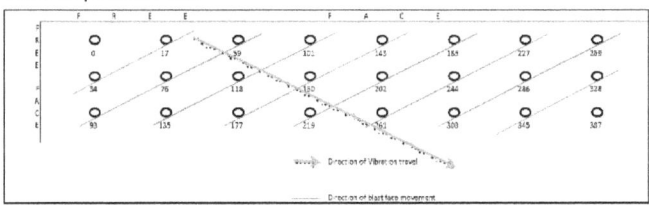

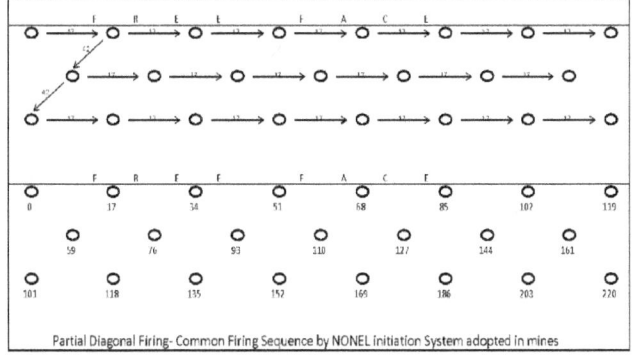

Partial Diagonal Firing- Common Firing Sequence by NONEL initiation System adopted in mines

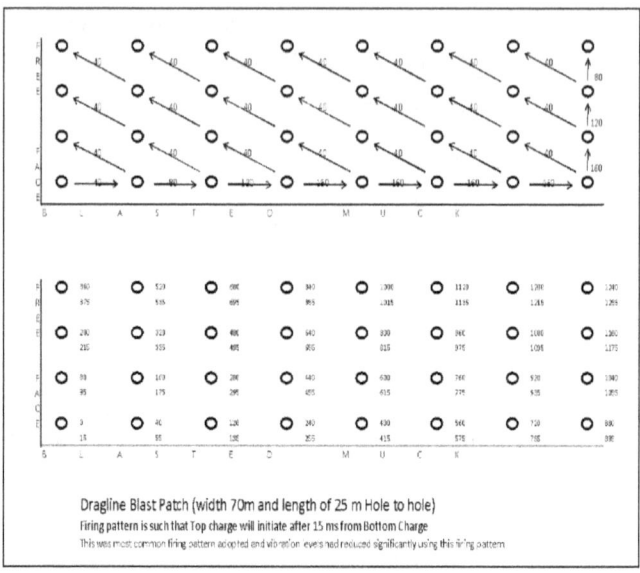

Dragline Blast Patch (width 70m and length of 25 m Hole to hole)

Firing pattern is such that Top charge will initiate after 15 ms from Bottom Charge

This was most common firing pattern adopted and vibration eves had reduced significantly using this firing pattern

3) V firing and extended V firing pattern:

This firing pattern is shown in the figure below. In This firing pattern, we get the benefits of diagonal firing but with a single free face. Burden is reduced by nearly 30%, hence fragmentation is better.

Another advantage with the V firing pattern is there is further breakage of rocks due to collision of rocks as there are two line of face breakage.

One issue with this firing pattern is that there can be overlapping of delay time between holes, though they will be certain distance apart, that will not impact fragmentation, but in this case, due to increase in maximum charge per delay, there could be increased chances of vibration.

One added advantage of this firing pattern is that the ground vibration produced during blasting, the direction of blast vibration line is distributed into 2 directions, due to two different blast faces generated.

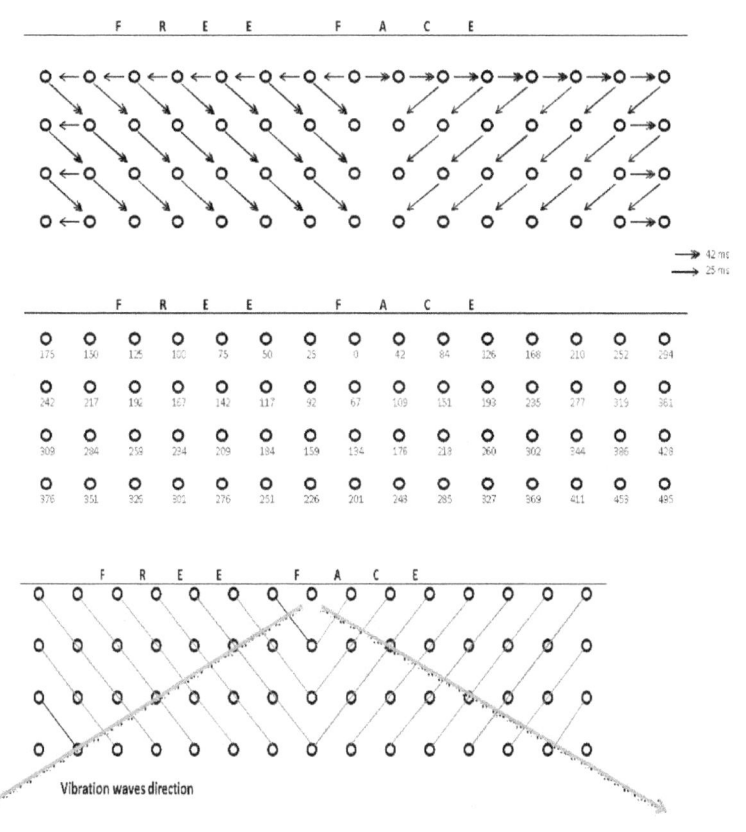

4) Special firing pattern adopted in special cases like Multi
   seam blasting, Blasting with no free face

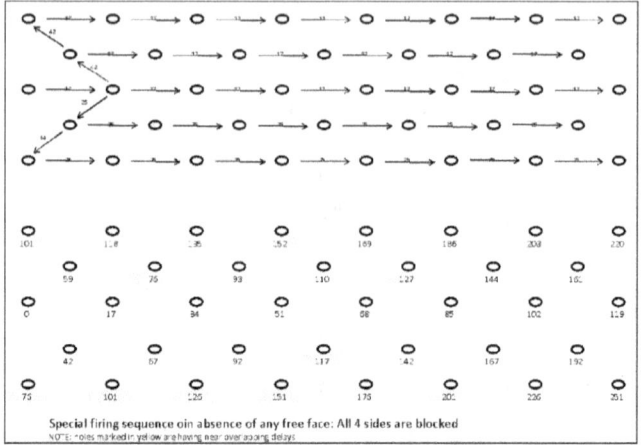

Consider the example of dragline delay balancing using MS connectors of 25ms and 42 ms and 67ms in fig below

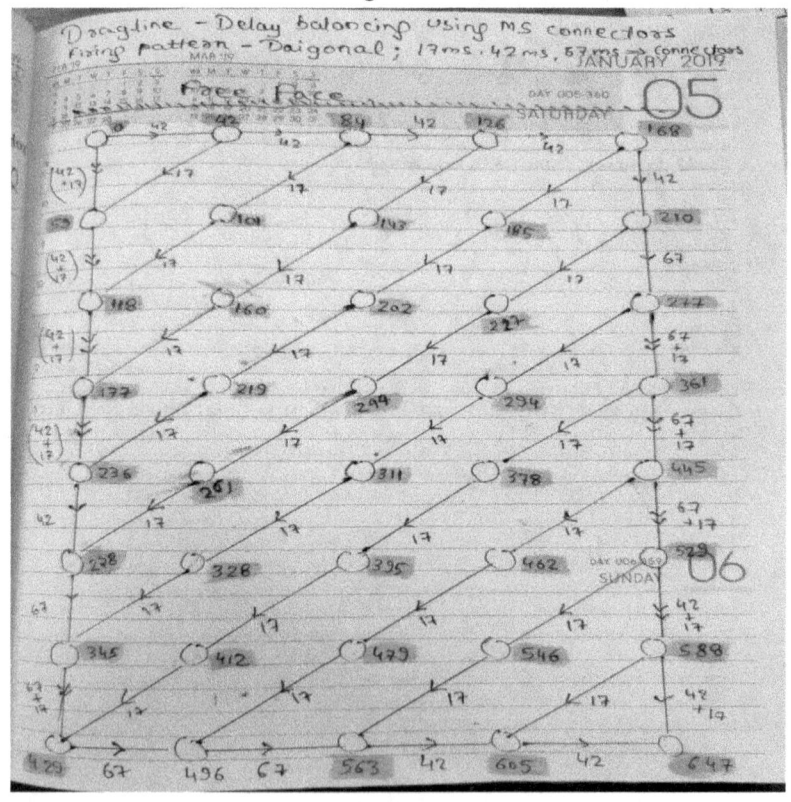

5) Segregation blasting: This blasting technique is used when the ore body is in contact with overburden rock (low grade) and the blasting is done in a manner that the muck pile of ore and waste rock is formed differently does not get mixed. This prevents dilution of ore and hence reduces handling and milling cost.

SAFETY IN FIELD OPERATIONS

Safety is of utmost importance during handling of any explosives.

Safe handling of explosive: Friction, Impact, Static charge and Heat (FISH) analysis for all explosive products used in field operations

Friction:

For testing of friction sensitivity, BAM machine is used. It has a base plate which is fixed and movable plate of porcelain.
Primary charge that is present in detonators is very sensitive and can be easily initiated by friction. PETN has higher resistance to handle friction but unnecessary mis-handling can initiate the PETN in cast booster or (as base charge in detonators also)
Bulk Explosive do not respond to or get initiated by friction yet if it gets very high temperature it theoretically it should get initiated.
Bulk Matrix do not blast (before adding gas bubbles)

Friction sensitivity as per test for Primary charge like lead azide is 0.02 Kg f and for base charge of PETN this value is 92 Kg f which is relatively safer than Lead Azide.

A: Initiating System:
1) NONEL: Highly sensitive to friction and must be handled carefully.
2) Detonating Fuse: Relatively less sensitive as compared to detonators, but should be handled carefully.
3) Electronic detonator: Highly sensitive to friction and must be handled carefully.

B: Accessories
4) Electric detonators: Highly sensitive to friction and must be handled carefully.
5) Cord Relays: Highly sensitive to friction and must be handled carefully.
6) MS Connectors: Highly sensitive to friction and must be handled carefully.

C: Primer Charge
1) Cast Booster: Made from PETN and TNT base. Must be safely handled.
2) Emulsion Booster: These contain paint grade Aluminium powder for increasing sensitivity, hence they are not as safe as emulsion. They also can be initiated with friction
3) Prime Cartridge: These contain paint grade Aluminium powder for increasing sensitivity, hence they are not as safe as emulsion. They also can be initiated with friction

D: Bulk Explosive/Base charge
These are safe and have high friction resistance. If subjected to friction the emulsion breaks and water and oil may separate.

Measures to prevent detonation of explosive due to friction:
All detonators must be handled carefully. There are separated bags (leather) or wooden boxes used for carrying the accessories.

Impact:

Drop weight Impact sensitivity test is done of all explosives. In this test a drop hammer of 2.5 kg is dropped on small sample and the height from which 50% sample get initiated, then that value gives impact sensitivity. Less the height of impact, less is impact resistance and easier to initiate the explosive.

PETN drop weight height is 35 cm, for lead azide it is 55 cm.

A: Initiating System:
1) NONEL: Highly sensitive to impact and must be handled carefully.
2) Detonating Fuse: Sensitive to impact and should be handled carefully.
3) Electronic detonator: Highly sensitive to impact and must be handled carefully.

B: Accessories
4) Electric detonators: Highly sensitive to impact and must be handled carefully.
5) Cord Relays: Highly sensitive to impact and must be handled carefully.
6) MS Connectors: Highly sensitive to impact and must be handled carefully.

C: Primer Charge
7) Cast Booster: Made from PETN and TNT base. Must be safely handled.
8) Emulsion Booster: These contain paint grade Aluminium powder for increasing sensitivity, hence they are not as safe as emulsion. They also can be initiated with impact
9) Prime Cartridge: These contain paint grade Aluminium powder for increasing sensitivity, hence they are not as safe as emulsion. They also can be initiated with impact

Measures to prevent detonation of explosive due to Impact:

All necessary steps must be taken to prevent any impact on detonator, boosters or prime explosive. At blast face all accessories must be kept with full care and should not be overrun by vehicles moving on blast patch.

During charging of the blast holes, the delivery hose must be lowered to the bottom of the blast hole during charging operations in dry or watery holes.

Static Charge:

Development of static charge is very common phenomena. The mechanism of initiation of any explosive material by static energy is caused in 3 steps.

A) Charging of capacitor to a potential. The capacitor can be 2 different plates or object carrying charge and separated by a small non conducting dielectric medium.

B) Capacitor must lose the charge due to break down of dielectric medium and there is generation of spark. When an electrically charged comes near opposite charged particles, to maintain charge balance, there is transfer of electrons which can create intense energy which appears in form of flash (due to breakdown of dielectric medium).

C) The path of this discharge must be sufficiently close to sensitive material to cause initiation.

A: Initiating System:

1) NONEL: Highly sensitive to Static charge as detonators contain primary explosive that are very sensitive and must be handled carefully.

2) Detonating Fuse: These have better static charge resistance as compared to primary charges but still can be initiated by sparks. But if there are thundering and

lightening, the Discharge potential is very high and should never be handled in such conditions.

3) Electronic detonator: As per the specifications, electronic detonators have more resistance to static charge as the charge gets dissipated via outer shell body to ground. But since it has a primary charge (ASA) which is very sensitive, one must not handle Electronic detonators during thundering.

B: Accessories

4) Electric detonators: Highly sensitive to Static charge as detonators contain primary explosive that are very sensitive and must be handled carefully.

5) Cord Relays: Highly sensitive to Static charge as detonators contain primary explosive that are very sensitive and must be handled carefully.MS Connectors: Highly sensitive to impact and must be handled carefully.

C: Primer Charge

1) Cast Booster: Resistance to small sparks due to ESD yet must never be handled during thundering or bad weather conditions.

2) Emulsion Booster: Resistance to small sparks due to ESD yet must never be handled during thundering or bad weather conditions. These contain paint grade Aluminium powder for increasing sensitivity, hence they are not as safe as emulsion.

3) Prime Cartridge: These contain paint grade Aluminium powder for increasing sensitivity, hence they are not as safe as emulsion.

D: Bulk Explosive: These do not initiate because of spark, ESD (not sure with lightening), heat or impact but these can be source of Static charge that can be generated because of friction between moving rotor and bulk, during pumping of bulk and it moves through hose etc.

Measures to prevent detonation of explosive due to Static charge:

This ESD has so much potential that it can easily fail any electronic circuitry and has led to number of accident in manufacturing industries. The voltage that can be generated just by human can range in thousand volts (for an instant). In BMD vehicles, there is provision of static charge dissipation to prevent any ESD as there are earth chains, which are grounded before any charging operation begins. Also the hose for delivery of bulk are anti-static. Anti static hose means that when explosive is passed through hose and if there is any static charge generated, then charge is dissipated to ground.

Thundering is Electro static discharge at very high potential. All explosives must be protected from thundering. Various risk related to handling of explosives are identified and ways to mitigate those risks at field.

Primary precautions that are taken are those related with handling of the accessories used in blasting.

Of the all accessories, the most sensitive is detonators be it electric detonators, electronic detonators or NONEL detonators. These detonators have primary charge and base charge that can be initiated by stray charges accumulation. Though electronic detonators manufacturers claim these to be safe from static electric charge, yet usage during thundering is strictly prohibited.

Even during thundering, any bulk charging operation need to be stopped.

Thundering or lightening contains huge energy and this energy can initiate bulk explosive, can charge detonators and initiate them and even can initiate detonating Fuse.

So provision for discharge of accumulated electric charge is very important. For this earth chain Anti static hose are provided.

Testing conductivity ensures the earth system is intact and working. To prevent accumulation of static charge, check the conductivity of chain links, delivery hose etc. this is done by a test instrument.

There can be provision of Installation of Copper cable along with the chain links as additional grounding chain in BMD vehicles. Person deployed in working in field must wear anti static safety shoes, cotton clothes, helmets, gloves, safety goggles, gum boots and raincoats for rainy season.

Heat:

Heat can initiate primary explosive and must be handled in safe ways. Heat sensitivity test is done for detonators and boosters. For cast booster, Heat sensitivity test is done by placing the booster for 24 hours at 75^0 C. PETN boosters when put is fire, burns releasing big yellow flames but did not detonate in open. Similarly the bulk explosive will release gases on heating but will not detonate.

A: Initiating System:
1) NONEL: Highly sensitive to heat and must not be primed in hot holes.
2) Detonating Fuse: Relatively less sensitive as compared to detonators, but should be handled carefully.
3) Electronic detonator: Highly sensitive to haet and must be handled carefully.

B: Accessories
4) Electric detonators: Highly sensitive to heat and must be handled carefully.
5) Cord Relays: Highly sensitive to heat and must be handled carefully.
6) MS Connectors: Highly sensitive to heat and must be handled carefully.

C: Primer Charge
4) Cast Booster: Made from PETN and TNT base. Must be safely handled.
5) Emulsion Booster: These contain paint grade Aluminium powder for increasing sensitivity, hence they are not as

safe as emulsion. They are safe from heat, but should not be charged in holes when temperature exceeds 80°C. These are relatively more resistant to heat as compared cast boosters.

6) Prime Cartridge: These contain paint grade Aluminium powder for increasing sensitivity, hence they are not as safe as emulsion. They also can be initiated with heat.

D: Bulk Explosive/Base charge
These have high thermal resistance. Charging of holes with temperature above 80°C is strictly prohibited.

Measures to prevent detonation of explosive due to Heat:
Special precautions are taken while charging of hot strata patches such as measuring the hole temperature and charging on when temperature of hole is below 80deg C. Charging of explosive in holes and after charging of holes priming of hole is done using Detonating Fuse and emulsion boosters.

One must have safety SOP/ guidelines duly prepared and readily available to the personnel involved in blasting operation. A general safety guidelines may include

a. Barricading the blast patch using tape and flags
b. Positioning the BMD away from holes
c. Prohibition of use of any electric or electronic item inside blast patch
d. Use of separate lane for Light Motor Vehicles (LMVs)
e. Proper care taken during priming of the blast holes
f. Use of wooden box/ leather bag for storage of electric detonators
g. Strict measures to prevent pilferage of detonators

h. During charging of explosive one must monitor running temperature and pressure in BMD system
i. PPE should be available
j. Training to be provided about general blasting and working in mines to all work force

Charging in Hot and cracked strata

Charging in cracked holes

Charging at cracked holes is a very big issue at many mines. In outsourcing benches, dealing with cracked holes are not much a big issue and providing cut bags of AN is sufficient to cater to this issue.

In cracked holes the most common solution is use of liners. Liner is thick polythene that is lowered in the blast hole and then explosive is charged. This prevents seepage of explosive outside the blast holes and thus improving blasting performance. But if the cracks are wide that then liner will not be able to bear load in higher depth blastholes. In such cases, decking of explosive using cut bags are done.

In deeper holes at shovel and Dragline bench this issue becomes severe.

Charging is done with cut bags, and if column doesn't go up then decking of hole is done. Many times crack is big and even during decking column of explosive doesn't rise. In that case the explosive decking is done using cut bags are by air filled bags. Using this technique column usually rises and issue is sorted out but if the issue of crack is still present then hole is first jammed using air bags and if not available then cut bags filled with drill cuttings are suspended and positioned in holes above crack and then top or remaining portion of hole is charged. The cut bags are suspended using 3-4 layers rope and tied on surface from large stone so that the load can be taken.

Charging at Hot holes

Presence of hot holes in the mines is also a potential hazard if proper planning is not done prior to charging of these holes. Due care is taken by during charging of these holes.

There are number of reason which causes increase in temperature of the holes and causes will require different methodology for handling situation.

1) Presence of a reactive ground: There could be presence of some reactive chemicals that may interact with Ammonium nitrate present in explosive and could increase the temperature and if exposed for long duration can cause initiation of explosive in the blast holes. Commonly pyritic ore have tendency of reacting and spontaneous combustion can take place.

2) Geo thermal causes: Another reason of presence of hot holes could be due to presence of some radioactive element present in the earth which raises temperature of the strata.

3) Combustion: In coal mines where old workings have been done the coal gets heated due to spontaneous combustion and strata nearby gets heated. This is very common reason of hot holes occurring in Coal mines.

Primary hazards that are associated with the charging of hot holes is that there are chances of premature detonation of explosives.

One more incident which has been experienced is that in very hot holes, water is filled and left for sometime to reduce the temperature but water starts boiling and hot steam is ejected from holes. This can lead to serious burn injuries. Many burn injuries have occurred due to hot holes while accessing the temperature.

Temperature of Blast holes can be measured using Infrared thermometers. If temperature of hole is more than 80^0 C, then hole is left uncharged.

Detonating Fuse is lowered in holes without any booster and checked if the fuse is burnt or not. If DF is burnt hole will not be charged.

Generally, the problem of higher blast-hole temperature is tackled by quenching with water. Proper quenching methods are to be applied to keep the temperature within 80 degree centigrade (as per DGMS criteria) . It has been observed that, flushing hot-holes with the mixture of water, bentolite, sodium silicate and guar gum solution help retain water to seal micro-fractures and cracks and bring the temperature down relatively easier. Holes with higher temperature should be identified and quenching of those hot-holes should be initiated at least 12 hours prior to charging and blasting in order to lower down the temperature.

When the temperature of hole is reduced, then explosive is charged in hole without priming of hole, so there is no risk of initiation of detonator or booster. After sometime (given for Gassing of Bulk Explosive) this charged hole is primed using DF and booster and small amount of explosive is again charged (that was kept beside hole during charging) so that booster is completely in contact with bulk. Stemming is completed and patch is blasted without much delay. In this case the primer is positioned at the top, hence an example of top priming of the charged column.

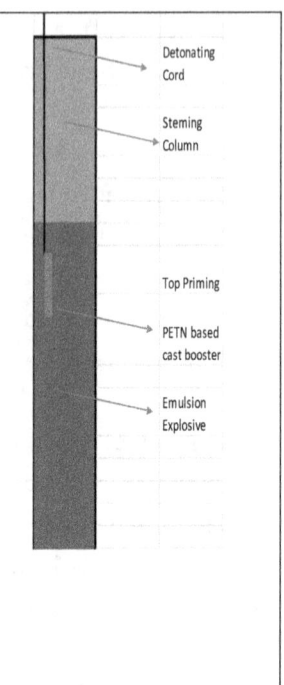

The entire patch is to be blasted in shortest time possible (within 2 hours from charging of first hole). Charging is done only when temperature is less than 80deg C.

Initiating system for blast initiation is Detonating fuse only and any other initiating system is not used (Shock Tube or detonators are susceptible to initiation due to heat).

If however the hole is still hot after filling it with water, then hole is left uncharged. These uncharged holes may lead to boulder generation, and hence pose huge distress post blasting during mucking operations.

For measuring temperature of hole, infrared thermometers are also available in mines.

Primary precaution to be taken is that during and post charging, minimum manpower is near blasting face and much time is not taken for blasting.

Hot water vapour that comes out from hot holes is very dangerous and can cause burn injury.

Charging at Dragline and deep hole shovel benches

Cast Blasting

Cast blasting is a way of blasting a bench such that blasted material directly gets thrown away and there is a minimum requirement of handling of this material. This is easily possible in dragline benches where blasted muck can directly be casted in a de-coaled area, and there is no handling of this casted muck.

This saves handling of blasted material, hence has many indirect cost advantages such as higher productivity, reduced cycle time, lower handling means lower cost of maintenance, faster rate of muck removal. But it has its disadvantages also. Quantity of explosive consumed in a blast will increase. To reduce this increased quantity, skipping of top charge in front rows and increasing bottom charge by 100-200 kgs for such holes can be done. This saved quantity of explosive could be used to cater more explosive demand in back rows. Since the boulders generated in front rows are not handled by any machine and will directly go into the de-coaled area, that was not an issue either. For effective cast blasting, drilling parameters should be such that there is good movement space available for front 3-5 rows. For this, bottom charge per holes was increased, and in front rows top charge was not given. Even in the second row and third row, the stemming column can be increased as compared with the last rows. Amount of primer cast booster is slightly less than 0.2% of the explosive charge i.e. if 1000 kg bulk is charged in the bottom, then nearly 1.75 kg-2.0 kg of cast booster is used. Burden for last rows needs to be less as compared with front rows so that effective push is provided by last rows of holes. Reduce stemming of last few rows of blast holes, as the fragmentation mainly is visible on top and in back, so first impression should be good. But this reduction of the stemming column must be done considering the distance of dragline and electric cables and bus bar boxes and other machinery so that there are minimum fly rocks possible.

Finally, the best delay pattern as per my experience is line firing or partial diagonal firing. 4-5 holes of the first row should be blasted and after that only the back row should blast. For top and bottom charge delay, there was in hole delay of 15-20 ms and hole to hole delay of 40-50 ms in the same row and row to row delay of 100 ms for the first 3 rows and that will increase to 150 to 200 ms in the back rows. Total delay for the blast should be restricted within 2000 ms (2 second).

Sometimes the mine management didn't want casting of blasted muck into the de-coaled area which is subjected to many practical requirements of the mine. These issues like pre-installed floating water pumps that cater to the supply of water in mines or there may be some ore left in old cut and haul roads are not to be blocked. In such cases the demand was also for near zero casting of blasted material.

In such instances following precautions can be taken

    1)   Reduce number of column

2) Firing pattern should cast material toward the previous blasted muck (diagonal fire)

3) Burden in front row to be increased and bottom and top charge in front rows to be reduced.

4) Increase in the number of decks to distribute explosive, hence decreasing energy concentration and thus reducing possibility of muck pile movement.

One important observation was that in cast blasting vibration was not very significantly high. And vibration in double deck charging at Dragline bench was also not very significant, despite the quantity of explosive per delay was high. This could be because if there is proper movement of burden, then maximum wave energy is dissipated and wave energy that moves in the form of vibration has lower impact.

Usual charging pattern for 30m holes with single deck and double deck is represented in fig. below.

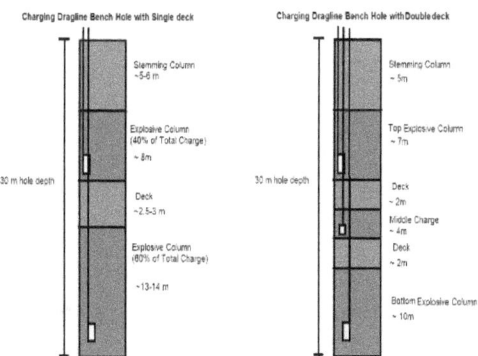

Toe formation in Dragline/shovel Bench

When the bottom portion of any blast is not efficiently fragmented and hard rock presence is seen then such a portion is called toe or elevated floors. In this case one has to opt for secondary holes drilling and blasting. But this secondary blasting is very cumbersome process, as it involves new face preparation, marching of drills, and usually shot holes are drilled. These short holes are charged but they impose a danger of flyrocks. There are many reasons for toe formation

A) The change in geological condition of strata is most common reason as well as scapegoat for explosive manufacturer and Shotfirer. Though it is an important factor and changing bed planes can lead to generation of toe. Presence of some intrusion in the seam, bedding of harder seam below. In case of dipping bed, if the blasting is along the strike, burden should be slightly reduced and if required spacing can be increased. Similarly if the direction of the blast is along the dip, spacing should slightly be reduced.

B) The method to correct the floor problems with this type of geological condition will depend upon the direction in which the blasting is progressing. If the blasting is progressing along the strike (parallel to the bed) the spacing between boreholes around this seam can be tightened up slightly. If the blasting is progressing up dip (perpendicular to the bed) the burden should be reduced when blasting in and behind this bed.

C) Irregular Hole depth is most common reason of toe formation

D) Bottom of hole has mud accumulated which is not settled and if explosive is charged in the hole, the explosive will mix with mud and effective blasting will not happen. Down the hole charging should be done in wet as well as dry holes.

E) Primer is positioned too much above from bottom

F) Density of explosive in the bottom of hole becomes more than critical density
G) Top initiation
H) Improper stemming that can cause release of energy without properly dislodging the bottom of the rock. Improper stemming results in decreasing gas retention time post blasting and is cause of many issues in blasting, including air overpressure and flyrocks

Problem faced during handling of toe:

A) Toe formation badly hampers production of Dragline and shovels, thus reducing productivity of these excavators
B) Since the drill has to be done in area that is inaccessible, there is associated cost of Dozer, drilling and face preparation
C) Safety of BMD vehicle is important as drilled patch is not very well connected.
D) Since the blast holes are of lesser depth, usually they are small holes, so the chances of fly rocks are very high. Also the patch may not be all hard face and there could be drilling in loose area, hence charging must be done taking extreme care.
E) One must try to doze off the hard area if possible using mechanical ripper dozer and secondary blasting must be last resort.

Prevention of Toe in dragline:
A) Regular depths of hole must be charged. Those holes which are not coal touch must not be charged.
B) Down the hole charging must be done and it must be ensured that the primer is in contact with the explosive. Time must be given after drilling of hole for settlement of mud (in case of wet hole).

C) It has been observed that there is significant reduction in toe formation with introduction of bottom initiating devices (NONEL and electronic detonators) as compared to Detonating fuse.

Primary reason for this is in bottom initiation, the gases generated from blast have more time for interaction with rock mass, hence proper movement of rock takes place. On the other hand in top initiation (as top is blasted first and bottom is blasted later with DF), the probability of gas leakage from cracks, vents are more thus leaving bottom portion hard sometimes.

D) Another important criteria is delay should be optimum. If the delay time for different rows is too less, then there will not be sufficient time for movement of rock, thus for holes in back rows, effective burden will increase and there will be chances of toe formation. On the contrary if the delay is too much, there will be more boulders as there will be leakage and dead pressing of explosive due to generation of cracks and shock from front rows. Also long delays will increase duration of blasting, thus vibration will sustain for longer duration.

E) Bottom density of explosive must be above critical density/dead press density. This can be checked using a density graph.

F) In the back side of the blast patch, some relieving holes are drilled. This is done to primarily increase charge concentration in the back rows, which will provide more energy for proper movement of muck pile. Another way to increase the charge concentration is by reducing the deck. Reducing burden is also common practice. This will reduce additional cost of drilling Pilot or relieving holes.

Multi seam Blasting

Introduction

The conventional practice followed in Opencast mines for extracting coal is blasting of OB for removal and thereafter blasting of coal. For this, the face is prepared for drilling and after drilling is complete holes are charged with initiators and explosive, which is done for each blast.

A new technique of blasting multi layered strata (OB and Coal) has been done in recent times. In this case, coal seam of some depth along with overburden drilled together has been blasted together, without intermixing of the two layers.

This technique eliminates the need for separate blasting of OB and coal, hence reducing the frequency of blast, and other cyclic works that follows. This also reduces downtime because of blasting and improves productivity

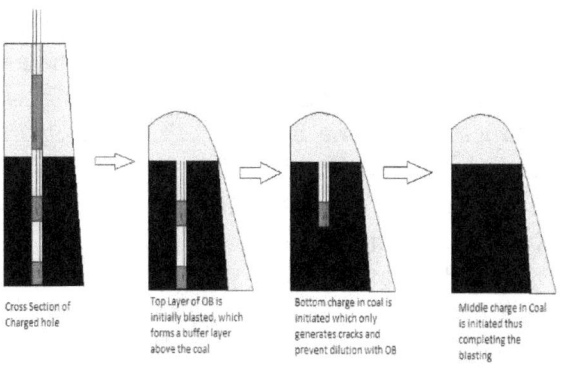

Cross Section of Charged hole

Top Layer of OB is initially blasted, which forms a buffer layer above the coal

Bottom charge in coal is initiated which only generates cracks and prevent dilution with OB

Middle charge in Coal is initiated thus completing the blasting

Minimum displacement in coal is required so that the dilution is minimized.

Primary requirement of the blast patch is ascertaining the location of various strata, so that proper positioning of decks and explosives could be achieved. Reduced or nil charges were employed where the blast hole intersects the two seams in order to reduce damage of coal seams.

As evident from the figure, the top layer of OB is blasted first and delay is arranged in such a way that there is good movement of OB.

The blast in coal is taken with care such that the blasted coal does not move and only gets loosened. This prevents dilution of coal.

Delay Timing for the top layer is set so that the fired shots are normally blasted as per the requirement of muck profile. Pre-split holes are fired initially at 0 ms and the first main hole is initiated at some 500 ms delay from pre-split holes. Delay provided between OB and coal is such that ample time is provided for settling of OB before coal is initiated. THis delay can be more than 3000 ms or more. Also the delay in coal or bottom seam is provided keeping in mind that no mixing or dilution of ore should occur.

Advantages

A) Blasting multiple layers in a single cycle. This alleviates the need of separate blast clean up, drill hole surveying, drill rig setup, explosive loading, blast firing sequence.
B) Reduces downtime because of blasting clearances and blasting operations
C) Each additional blast is source of dust, fumes, noise and other environmental issues
D) Reduced vibrations, noise, fly rocks

Disadvantage

1. The primary disadvantage is exact location of seam parting must be known
2. The fragmentation in the bottom seam will not be satisfactory and there may be some hard portion. This is

because caution is taken to prevent mixing of two strata, to prevent dilution.

Post Blast Analysis

Post Blast analysis

Blast performance is very important and is checked as a post blast result. Following parameters are checked for analysis of blast performance.

1: Profile of blasted muck

2: Fragment size and distribution

3: Cycle time of equipment engaged in mucking of blasted material.

4: Powder factor

5: Toe formation and incomplete lifting

Profile of blasted muck:

AS can be seen from the figure, a typical blasted muck profile has 3 main components. Throw, drop and lateral spread. For different excavators, the blast profile is different.

Diggability of the blasted muck depends upon the excavator used as well as profile of muck pile formed post blast. There must be optimum throw and drop for proper digging.

For the Shovel bench, lateral spread must be less and drop should be optimum, whereas for Dragline bench, Lateral spread and throw should be high, but drop should be optimum. If drop is more than required, then Dragline has to backfill some blasted material so as to create a stable base for sitting.

Similarly for the Backhoe, Lateral spread must be lower, drop must be lower and throw must be optimum.

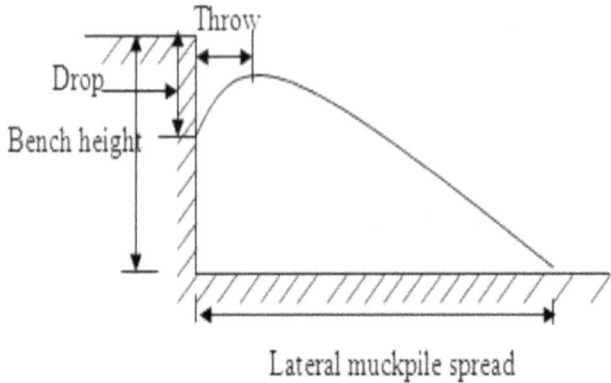

Lateral muckpile spread

Fig.1: Muckpile shape parameters

There are number of factors that determine the profile of blasted muck pile.

Increasing the burden reduces throw, lateral spread and drop also.

Muck profile depends upon number of blast design parameters, such as spacing, burden, drill hole diameter, depth of holes, bench stiffness ratio and powder factor.

Fragmentation:

This is the most important post blast result. Analysis of fragmentation post blast can be done using Digital photography and then analysing the fragment size using a reference length. There are number of software available that can be used for fragmentation analysis and generation of curve showing fragment size and other related parameters like mean average size, maximum size of boulders generated.

Also number of researches are being done for prediction of fragmentation using number of parameters like bench height, spacing, burden, stemming length. Hardness of rock mass, presence of discontinuities, explosive energy, decoupling charges etc.

Generation of boulders badly affects the production, increases the cycle time of machineries, increases downtime of machineries and thus reduces overall productivity of mines.

Too fine fragmentation is also undesirable, as in this process consumption of explosive is not optimised. Excess explosive puts additional cost to company.

Outcome of fragmentation depends on number of factors

A: Uncontrollable Factors: Geology of Deposit, rock strength, presence of discontinuities

B: Controllable Factors:

    Blast design parameters like diameter, charge length, burden, spacing, stemming column.

    Properties of Explosive used in blast.

    Delay time and firing sequence

Numbers of Software have been developed for analysis of fragmentation. These Software uses digital photographs of fragmented muckpile and provides fragment distribution, gives mean fragment size (K50) and maximum fragment size (K100).

Importance of Drill pattern and fragmentation and Powder factor and there relation

This is very important to consider for a blast patch, what is the optimum drill pattern that could be reliably detonated.

The most common issue which is more psychological is that when a blast fragmentation is not sufficient, the trend is to reduce the burden and spacing for the blast patch. At first instance one should try to increase explosive charge per hole, rather than reducing pattern.

This could be understood using simple calculation

Consider Case 1:

Burden= 4m

Spacing= 5m

Depth of Hole=5m

Charge= 45 kg

Powder Factor= 4*5*5/45= 2.22 m3/kg of explosive

Case 2:

If in previous case blast fragmentation is not as desired and we do not change the drill pattern, rather increase charge per hole to 50 kg

Powder factor= 4*5*5/50= 2.0 m3/kg

Case 3; In this case if we decrease the pattern by 0.3 m for burden and spacing

We have

Powder factor= 3.7*4.7*5/45= 1.93 m3/kg

So even a small decrease in reducing pattern is a loss because that will not only decrease the blasting efficiency of explosives but also add the cost of drilling, which must be considered. THis is a very critical issue one faces when dealing with the management of Explosives.

One important fact is also there that there may be a case that the level of fragmentation in case 1 and Case3 remain the same, as there is no reduction of the stemming column. So slight reduction of the stemming column is the easiest way to improve blasting in respect to fragmentation. Reducing the pattern is the ultimate option available with the management, because the financial impact are huge.

Improper Lifting

Case 1: It has been commonly observed at shovel bench that if there is some hard toe left out in the bottom, the shovel tends to move upward at some inclination, hence effectively some loose material is left out. It also reduces bench height.

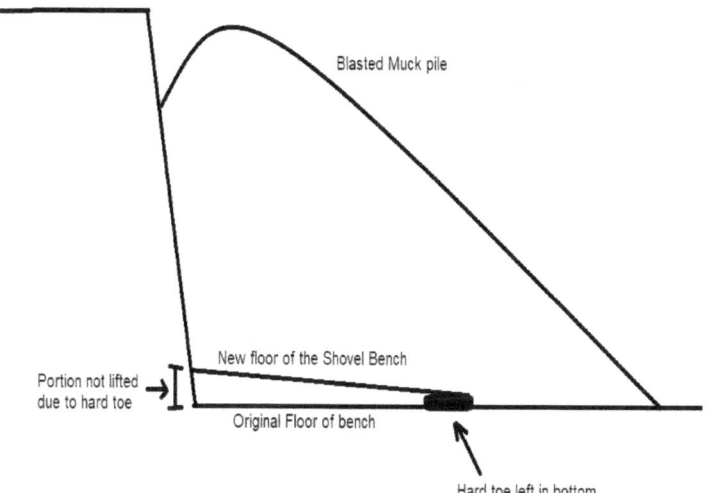

Blasted Muck pile

New floor of the Shovel Bench

Portion not lifted due to hard toe →

Original Floor of bench

Hard toe left in bottom

To cater this issue we need to doze off that hard toe and lift the loose material. This issue is very prevalent in deep shovel benches.

Multiple Choice Questions

Q1. During Blasting of rock

1: Maximum fragmentation occur in tension

2: Compressive strength of rock is much higher than tensile strength

Which of the following statement is right

| A: 1 & 2 are correct and 2 is correct explanation of 1 | C: 1 is correct and 2 is wrong |
| B: Both correct but 2 is not explanation of 1 | D: 2 is correct and 1 is wrong |

ANs: A

Q2. Which statement is correct

1: Explosive on blasting releases energy in 2 forms- Shock Energy and Gas energy

2: Shock Energy creates the cracks and gas energy causes the heaving of muck pile

3: Shock energy is initially a compressive waves, which upon reflection from free face gets phase change and is converted in tensile waves

4: Maximum work is done by gas energy

Options

| A: All correct | C: 2, 3, 4 correct |
| B: 1, 2, 3 correct | D: 1, 3, 4 correct |

Ans: A

Q3. In case of rock strata having plastic nature which statement is correct?

1: Gas energy gets easily attenuated and maximum work is done by shock waves generated

2: Shock waves travels large distances creating huge cracks and rock breaks due to spalling

3: Ground Vibration is lower

A: All correct	C: 2 correct
B: 1 correct	D: 3 correct

Ans 3

Q4. Poisson's Ratio is

1: Negative of the ratio of axial strain to transverse strain

2: It suggests the brittleness of the rocks, and required during blast design

Which of the statement is correct?

A: 1 & 2 are correct and 2 is correct explanation of 1	C: 1 is correct and 2 is wrong
B: Both correct but 2 is not explanation of 1	D: 2 is correct and 1 is wrong

ANs: D

Q5. Impact of presence of discontinuity in rock mass
1. Fault planes act as medium for release of gas energy thus performance of blast is hampered
2. Fault planes act as free face for the shock waves causing reflection of compressive waves, causing improper crack generation and affecting blast performance.
3. Profile formed in blasting has poor uniformity index.

Which of the following statement is false?

A: None of the above	C: 2
B: 1	D: 3

Ans: a

Q6: Impact of presence of cavity in rockmass

1. Fault planes act as medium for release of gas energy thus performance of blast is hampered
2. Fault planes act as free face for the shock waves causing reflection of compressive waves, causing improper crack generation and affecting blast performance.
3. Profile formed in blasting has poor uniformity index.
4. Which of the following statement is false?

| A: None of the above | C: 2 |
| B: 1 | D: 3 |

Ans: a

Q7: What is the way to improve blasting performance if discontinuity is present in rock mass?

| A: Decking the portion where discontinuity is present | C: Use of liners and cut bags to bring the charge column for desired stemming |
| B: Increase the number of holes by reducing drill pattern | D: All of the above |

Ans: 4

Q8: Consider following statement
1. ANFO is highly resistant to water
2. Ammonium nitrate is soluble in water

Which of the following statement is true?

| A: 1 | C: Both |
| B: 2 | D: None |

Q9: Which of the following statement is false?

| A: Leaching of emulsion explosive occurs due to presence of water in the strata | C: Strata water in motion can cause washout of explosive charge, thus can decrease efficiency of blast |

B: Rate of leaching is slow provided that the water is not in motion	D: Presence of water in blasthole reduces chances of stemming ejection

Ans: 4

Q10: Which of the following statement is correct?

A: Presence of reactive ground causes self heating and can be reason of initiation of explosive	C: In case of reactive strata, charging must be done as considering it hot strata and due precautions must be taken
B: Some sulphides and other ore have tendency to self heat and is reason for reactive strata	D: All of the above

Ans: 4

Q11: What is the sequence of rock fragmentation
1. Heaving of the rocks and formation of muck profile due to gas pressure
2. Formation of gases as a result of explosive reaction
3. Crushing of the rocks surrounding the hole due to intense compressive shock wave
4. Generation of shock energy and that moves is form of shock waves
5. Reflection of compressive waves from free face and turning into tensile waves
6. Breakage of rock in form of spalling and generation of cracks and fractures in the rock

A: 1-2-3-4-5-6	C: 2-1-4-3-5-6
B: 4-3-5-6-2-1	D: 4-5-6-3-1-2

Ans: b

Q12. Bench height is doesn't dependent on which of the parameters

A: Size and reach of the excavator	C: Quantity of explosive
B: Thickness of Strata and stratification	D: None of the above

Ans: 3

Q.13 Which statement is true regarding borehole diameter

A: VOD of explosive increases as confining diameter increases until steady state VOD is attained	C: Energy distribution is higher diameter blasthole is better and hence higher drill pattern is required
B: For deep hole blasting, diameter of hole should also increase	D: All of the above

Ans: d

Q.14 Usually the fragmentation in diagonal firing sequence is better than line firing delay sequence because-

1. Effective burden and spacing are reduced
2. In diagonal firing better movement of blast material

Which statements are correct

A: Both are correct	C: 2
B: 1	D: None

Q. 15 Sub grade drilling is required because

A: To prevent fly rocks in the blast	C: To prevent air over pressure
B: To prevent formation of toe in bottom of the bench	D: Not required and undesirable

Ans: 2

Q.16 What is difference between Hole depth and bench height

A: 0	C: Burden
B: Sub grade	D: No relation

Ans: B

Q17. What are uses of deck charging?

A: Helps to fill the soft strata, discontinuity with non explosive material and thus promotes efficient energy utilization.	C: Require more explosive charge per hole, thus increasing available energy
B: Reduces maximum charge per delay, thus increasing fly rocks	D: All the above

Ans: 1

Q18. Which of the following statement are true about Powder factor for explosive?

1. It is the amount of rock blasted per kilogram of explosive
2. It is only dependent on quality of explosive and fixed for a given strata or mines

Options

A: Both are correct	C: 2 is correct
B: 1 is correct	D: None

Ans: b

Q19: Numerical 1: Consider a situation where a blast has to be designed. The length of blast patch should be 40m and width of patch to be 20m. Bench height required is 5 m and sub grade is 10% of bench height. Powder factor to be achieved is 1.6 m3/kg of explosive. Loading density for the explosive charge is 25kg/m. Burden to spacing ratio is to be maintained in between at 1:1.2

to 1:1.3 and final stemming column is 3m, which of the following options are correct?

1: What is the depth of hole

A: 5m	C: 6.0 m
B: 5.5m	D: 4.5m

Ans: B (Total hole is bench height + subgrade)

2: Quantity of Explosive required per hole is

A: 50kg	C: 75 kgs
B: 62.5kg	D: 40 kgs

Ans: B

Total Stemming=3

Charge column =5.5-3= 2.5 m

Loading density @25kg/m, total charge = 25*2.5=62.5 kgs

3: Total number of holes to be drilled in blast patch

A: 30	C: 40
B: 35	D: 45

Ans: C

Total volume of block= 40*20*5= 4000 m^3

Target Powder Factor= 1.6m^3/kg

Total permitted explosive= 4000/1.6= 2500 kgs

Charge per hole=62.5 kg

Total permitted holes= 2500/62.5 =40

4: Burden and spacing for the blast patch is

A: 4* 5	C: 4*4.5
B: 3.5*4.2	D: 4.5*4.5

Ans: A

Total holes=40

Total Area to be blasted= 40*20= 800m^2

Area blasted by 1hole= 800/40= 20m^2

Desired pattern= 4*5 (also burden : spacing is 1:1.25)

Q20: Numerical 2:
Consider that a deep hole blasting of 30m bench height is required to be done for Dragline. Length of Sub grade drilling required for the blasting patch is 2m and it has been decided that to reduce maximum charge per delay, one deck will be placed that will divide entire explosive charge such that 60% of the total charge is bottom charge and 40% is top charge. Loading density of explosive is 70kg/m. The stemming height left is 5m and deck is 3m. Reach of dragline boom is 80m hence that is width of blast patch and length to width ratio is 2:1 for proper casting. Also required Powder Factor is 1.5m3/kg of explosive, determine the following:

1. Total explosive charge per hole

A: 1400 kg	C: 1680 kg
B: 1540 kg	D: 1820 kg

Ans: C

Total Charge length= 30+2-5-3=24 m
Loading density= 70 kg/m
Total charge per hole= 24*70= 1680 kg

2. Total volume of rock blasted in entire patch (m3)

A: 384000	C: 96000
B: 409600	D: 102400

Ans: A

Width of Blast patch = 80 m
Length of blast patch= 160 m
Bench Height= 30 m
Total volume= 80*160*30= 384000

3. What is the bottom explosive charge required

A: 1000 kg	C: 1100 kg

B: 900 kg	D: 1200 kg

Ans: A

Total charge per hole= 1680 kg

Bottom charge is 60% of total charge, so bottom charge = 0.6*1680= 1008 kg (nearly 1000 kg)

4. What is quantity required in top charge

A: 680 kg	C: 720 kg
B: 640 kg	D: 600 kg

Ans: A

5. Total number of holes required in blast patch

A: 152	C: 144
B: 168	D: 175

Ans: A

Total Volume =384000

Desired Powder factor= $1.5m^3/kg$

Total Explosives= 256000 kg

Charge per hole= 1680 kg

Number of Holes= 152 (nearly)

6. What is tentative drilling pattern required for the above blast patch (Burden * Spacing in m)

A: 9* 11	C: 8.5*10
B: 9*10	D: 8.5*11

Ans: C

Total area of blast patch= 80*160= $12800m^2$

Total number of holes= 152

Average area per hole: 84 m^2

So tentative drill pattern can be 8.5* 10

Q21: What is bench stiffness ratio?

A: Ratio of bench height to spacing	C: Ratio of bench height to burden
B: Ratio of hole depth to sub grade dilling	D: Ratio of Hole depth to bench height

Ans: C

Q22: Which of the following statement is not correct for Line drilling?

1. It helps in controlling Ground Vibration
2. It helps in controlling fly rocks
3. It provides good high wall post blasting
4. It prevent generation of back breaks

Option

A: 1	D: 4
B:2	E: None
C:3	

Ans: B

Q23: Which of the following is not an initiating system?

A: Shock Tube	C: Detonating Fuse
B: Electronic detonator	D: Cord relay

Ans: D

Q24: What is Primer?

A: It is explosive unit which provides energy required for initiation	C: It is the system of transmission of energy to each blast hole for initiation of explosive
B: It is explosive unit attached with detonator for initiation of explosive	D: It is same as cast booster

Ans: B

Q25: Which of the following is not correct about Detonating fuse?

A: It contain layer of PETN inside	C: Delays are provided in DF using MS Connectors and shock tubes
B: It deflagrates as it blast	D: It provides sufficient energy to initiate cap sensitive explosives

Ans: B & C

Q26: Which of the statement is correct about DF?

A: There is noise in DF as it detonates as waves travel along the tube	C: Chances of stemming ejection increases in case of DF as initiating system
B: DF causes loosening of stemming, thus reducing gas retention time, and performance is better as compared to other initiation system in small diameter holes	D: Changes of Toe formation in high depth hole blasting are less as compared to other initiation system.

Ans: 1 and 3

Q27: Using Detonating Fuse as initiating system, statements are listed in comparison with NONEL as Initiating system. Point out the demerits

1. Gas retention time is lower
2. Stemming Loosening in case of low diameter holes
3. De-sensitization of column of explosive charge
4. Top initiation
5. Non accurate delay timing
6. Easy to use
7. Higher chances of stemming ejection
8. Less sensitive to the shock and other hazards

Ans: 1, 2,3, 4, 5, 7

Q28: Which chemical is present as explosive inside Detonating fuse?

A: PETN	C: Lead Azide
B: Ammonium Nitrate	D: Nitroglycerin

Ans: 1

Q29: What is VOD range for Detonating Fuse?

A: 1500-2500 m/s	C: 4500-5500 m/s
B: 2500-3500 m/s	D: 6500-7500 m/s

Ans: 4

Q30: What explosive chemicals does Detonator contain?

A: PETN	C: Lead Styphnate
B: Lead Azide	D: Aluminium Powder

Ans: 1, 2, 3, 4

Q31: Which of the following is not present in Electronic Detonator?

A: Base charge	C: Delay element
B: Fusehead	D: Primary charge

Ans: 3

Q32: Which of the following is not present in NONEL Detonator?

A: Base charge	C: Delay element
B: Fusehead	D: Primary charge

Ans: 2

Q33: What is #8 strength detonator?

A: Detonator containing 2gm mixture of 80% Mercury Fulminate and 20% potassium chlorate	C: 0.4gm to 0.45 gm PETN as base charge
B: Detonator having energy	D: All of these

equivalent to above mentioned mixture	

Ans: 4

Q34: Overlapping of delay sequencing not lead to which of the following condition

A: Fly rocks	C: Increased Vibration
B: Back Breaks	D: Improper drill pattern

Ans: 4

Q35: In case of hot hole, consider the following statements
1. Detonating Fuse must be used in hot strata
2. Presence of confined primary charges can initiate detonators

Option

A: Both are correct and 2 is correct explanation of 1	C: 1 is wrong and 2 is correct
B: 1 is correct and 2 is wrong	D: Both are incorrect

Ans: A

Q36: Advantages of Electronic detonator over NONEL initiation system are

A: Bottom Initiation	C: Safe from extrinsic stray charges
B: Delay Precision	D: Reduced Back breaks

Ans: B and D

Q37: Which component is not available in Electronic Detonator?

A: Leg wire	C: ASA Primary charge
B: Base Charge as PETN	D: Lead filled pyrotechnic element for Delay

Ans: D

Q38: Which of the following is correct about Safety Fuse?

Q45: Which among the given is most sensitive in respect to initiation?

| A: Lead Azide | C: Bulk Emulsion Explosive |
| B: PETN | D: Cartridge Explosive |

Ans: 1

Q46: While charging in cracked holes special precaution that needs to be taken are?

| A: Use of Liners/ cutbags | C: Use of air bags for while charging bulk explosive |
| B: Decking in the portion where cracks are found | D: All the above |

Ans: 4

Q47: For charging of Hot holes consider following statements
1. Temperature of Holes must be below 80deg
2. Temperature of hole can be controlled by filling the holes with water
3. Detonating fuse not to be used for priming of hot holes
4. Charging and blasting of hot holes must be completed within 2.5 hours as per DGMS.

Options that are correct

| A: 1 and 2 | B: 1, 2 and 3 |
| C: 2 and 3 | D: 2 and 4 |

Ans: A

Q48: Consider following statement regarding charging of hot holes in mines
1. Infrared thermometers are used for measuring the hole temperature
2. Shock tubes can be used for blasting hot holes

3. Electronic detonators are most suitable initiating system for charging of the blast holes

Which of the following is correct

A: 1	C:3
B: 2	D: None

Ans: 1

Q49: What could be probable reason for hot holes during charging operations

A: Presence of reactive strata below such as coal seam, pyrites etc	C: Geothermal processes that may increase temp
B: Old working mines that were previously worked out using underground methods	D: All of the above

Ans: D

Q50: Consider the statement regarding cast blasting

    A. Quantity of explosive charge is higher as compared to normal blasts

    B. Vibration is considerably higher as compared to normal non cast blasts due to higher charge

Option

A: 1 and 2 are correct and 2 is correct explanation of 1	C: 1 is wrong
B: Both correct but 2 is not correct explanation of 1	D: 2 is wrong
E: Both are wrong	

Ans: D

Q51: Which among the following deviation will not be a reason for toe formation?

    1. Irregular hole depth in the blast face

    2. Position of booster in the charge column

3. Downhole delay of the blast hole
4. Detonating Fuse is used as initiating system
5. Presence of unsettled mud in the bottom oof holes

Options

A: 1	B: 2
C: 3	D: 4
E: 5	F: None of the above

Ans: F, all the above reason can lead to toe formation

Q 52: Issues while dealing with blasting of elevated floors/toes are?
1. Drill face is easily created
2. There is loss of production as this leads to downtime
3. Blast hole that are drilled have improper face and there are chances of fly rocks
4. Since the face condition is not good, these may pose some accidents to Machineries involved

Which if the option is incorrect

A: 1	B: 1,2,3,4
C: 2, 3, 4	D: 1 and 2

Ans: A

Q53: In case of Bottom initiation, the chances of Toe is

A: Increased as top portion is displaced later	B: Reduced as gas retention time is more
C: No such impact	D: Increased as gas retention time increases

Ans: B

Q54:

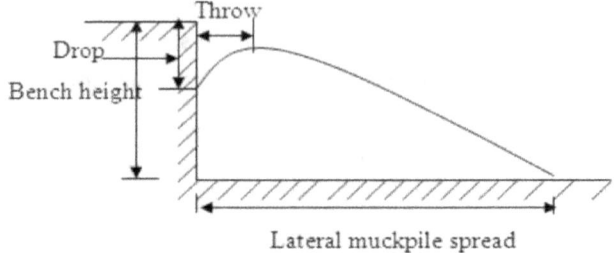

Fig.1: Muckpile shape parameters

Consider the Muck pile figure

Which is optimum muck pile for Shovel Bench

1. High Throw
2. High Lateral Spread
3. High Drop
4. None of the above

Ans: 4

Which is optimum muck pile for Dragline?

1. High Throw
2. High Lateral Spread
3. High Drop
4. None of the above

Ans: 1 and 2

What is optimum muck pile for LHD loaders

1. High Throw
2. High Lateral Spread
3. High Drop
4. None of the above

Ans: 1, 2 and 3

Chapter 4: Issues in blasting and measures to control

Various issues in blasting that are discussed are Fly Rocks, Poor Fragmentation, Vibration and Air overpressure, Blown outs/Stemming Ejection, Crystallisation of Bulk explosives, Issues with Low Viscosity product, Premature initiation of Electronic Detonator at NAKC

Stemming ejection or blowout.

Stemming is a process applied to blast holes to prevent gases from escaping during detonation. A stemming material helps confine the explosive energy for a longer duration.

Stemming ejection/ blown out of hole means the energy of gases that are generated during explosive is leaked out from top after removing the stem. This primarily happen when the stemming is not compact enough.

90% blasting failure is only one or other form of blowout. Improper stemming in watery holes is the biggest reason in this context. If stemming material is lighter and doesn't sink easily, then there are huge chances of blowout. Even tamping of the watery holes may cause jamming of stemming column, thus causing stemming blow out.

It has been tried that in case of water stemming the result of blast is very good.

When a watery hole is left as it is then also blasting will be good and if stemming is done properly then also blasting will be good, but improper stemming, tamping the holes or preventing settling of stemming lead to blowout of stemming material.

It has been observed that when crystallisation point of emulsion matrix is reduced, there will be an increase in cases of stemming ejection. I think these factors lowers the VOD of explosive, hence chances of stemming blowout increases.

Stemming blown outs are significantly reduced using NONEL or Electronic detonators as these will not disturb the stemming column and initiation of explosive is from bottom, hence more time is available for gas energy for movement of blasted material. Even in top initiation, chances of stemming blown outs can increase. Using Detonating Fuse as initiating system, as the fuse detonates as it propagates the blast it firstly looses the stemming column and also burn the explosive (Desensitize by bursting the gas bubbles) that surrounds DF.

There can be so many reasons of Stemming ejection

1. Loose stemming: It is good practice to maintain the stemming column near to the burden of the blast patch, which simply reduces chances of gas energy finding easy pass through Stemming. So stemming plays a very critical role in blasting. If stemming is high, there will be generation of boulders, if low, fly rocks, if loose blown outs. Optimum window for good stemming is very small.

2. High Burden or spacing: Again chances of blown out or stemming ejection increases, if the burden required to be broken is too high. Here the burden actually means effective burden. There may be the case that Burden was good but scatter in delay during blasting can cause increase in effective burden, so delay sequencing play a role in poor blast related to stemming ejection.

3. Watery holes, where stemming not settled properly
4. Improper firing sequence

5. High explosive charge column thus reducing the column of stemming, can result in blowout of holes.
6. Use of Detonating Fuse as initiation system, especially in small diameter holes: Use of detonating fuse causes the loosening of the stemming column because of its detonation. This can create a gap in stemming, reducing the gas retention, and thus creating inefficient blast. Though many blasts have been conducted using DF, cases of stemming ejection in NONEL or electronic detonators are far lower in comparison with DF as an initiating system.

7. Quality of Primer and Bulk Explosive: The chances of stemming ejection increases especially in the case of watery holes. Probable reason could be related to low VOD.

Undercharging:

There are instances when blasting technicians encounter new strata and are unaware of the hardness and strength of rocks, there are chances of undercharging which causes formation of boulders and only generation of cracks in the blasting patch. But this is not only the case why boulders are generated.

Number of uncontrollable and controllable factors that can lead to generation of boulders.

1. Uncontrollable Factors
   a. Presence of fold, fault, variation in strata, or other discontinuity
   b. Presence of hard strata in stemming column and softer strata in explosive charge column
   c. Presence of cracks and cavities
2. Controllable factors
   a. Improper inspection of the blasting patch
   b. Poor quality of Primer explosive
   c. Lower strength of Bulk Explosives

d. Human/Mechanical error during charging of Blast holes
e. It is common practice to control total explosive charge in the first row holes primarily to restrict the fly rocks that may be generated. But this practice leads to generation of boulders as the stemming column is increased. To prevent such happening, one should maintain a good front row burden and explosive charge must not be highly restricted.

Handling of Boulder:

Rock Breaker:
These are the hydraulic machines with hammer in front, which have been deployed for breaking of oversized boulders generated in blast.

Secondary Blasting
A. Pop Shooting
B. Mud capping

Drilling small diameter Holes in rock and charging it using a cartridge.

Making an small hole using normal Drill (same diameter as hole) and charging using SME

Taking SME explosives in bags and keeping that on the boulder to be blasted and plastered with mud also called capping of explosives. In this case the air overpressure generated is huge.

Disadvantages of Secondary Blasting
1) Additional requirement of small diameter drill machines for drilling purposes which beside investment cost has cost of wearable, operating cost of manpower and fuel.
2) Risk of Fly rocks and air overpressure as the stemming of these is not very compact. Also charge put in the

holes must be optimum and requires trained and experienced manpower.

3) Due to boulders, the efficiency or productivity of the excavator is sufficiently reduced.

4) Also blasting small charged holes requires additional muffling which is added work for the manpower

Over charging :

This is again a big issue if hole has been overcharged. This will lead to generation of fly rocks. If the drill pattern is lower than required, then also it could be overcharged and it may lead to generation of fly rocks. If the front burden is very less, extra care needs to be taken for charging front row holes as that will cause fly rocks in horizontal direction, and can hit machines or even people if present in front of a free face. Overcharging of a blast hole is far more problematic than an undercharged hole, as the former can lead to serious injuries.

The most common issue which can cause overcharging is lack of proper communication between the operator of BMD vehicle and the person measuring the depth of hole.

If it has come to knowledge that hole has been overcharged, one must leave that hole and not blast that specific hole. That hole must be treated as misfire and handled accordingly.

Ground Vibration:

Issue of ground vibration is a book in itself. This topic is most researched in opencast mining and blasting in particular. What makes this issue so important is that people living nearby mines are badly affected by ground vibration.

Pressure on mine management due to ground vibration is so huge that many times new technology is adapted to skip blasting in sensitive zones. Use of eccentric Ripper, using High wall mining and using the chemicals to break the boulders generated instead of doing secondary blasting are a few among many new technologies that are adopted just to skip blasting in sensitive zones.

Ground vibration has become a social and political issue. If a house is not properly constructed near the mine area there are huge chances that house may collapse within a few decades.

There are so many controllable and uncontrollable factors that are responsible for ground vibration. Rather as a matter of fact whatever we think related to blasting is a factor controlling vibration.

Vibration is to and fro movement of particles. Net displacement is zero but this vibration damages many buildings, zero displacement and crores of economic damage is done. Can we control ground vibration?

We can minimize it but making it zero is not possible at least using current methods of blasting.

Whenever we hit a solid surface then the particles start vibrating, this vibration is transmitted to other particles which are in contact, thus causing a wave like motion of energy moving through the particles. Same is the case with ground vibration as when blasting takes place, a part of energy moves toward highwall (Solid ground). The wave is formed in which particles just oscillates along neutral point and energy moves forward.

Displacement of particle can be expressed as sine function

$D = A * \sin(wt)$; where D is displacement , A is amplitude w is angular frequency and t is time.

$W = 2 * pi/T$ and T is time period of oscillation. Since frequency $f = 1/T$ so

$D = A * \sin(2 * pi * f * t)$

And velocity of particle at any instant is differential of D with time t, so

Particle velocity, $V = A * 2 * pi * f * \cos(wt)$ or $= Vmax * \cos(wt)$

Where V max=2*pi*f*A

Simple concept is more the displacement higher is the risk of structure failure. At mines PPV or peak particle velocity is the primary measured thing. So for given PPV, displacement and frequency are inversely related. Hence if frequency is high, displacement is low, hence this situation is okay (as displacement is primary cause of cracks) and low frequency and high PPV is never desired as displacement of particles will be huge and more impact will be felt.

There is an instrument used for measuring ground vibration called seismographs. They have tri-axial sensors, aligned on three axes and as these sensors move, vibration is recorded. These seismographs must be calibrated (nearly in 1 year), for accurate measure of ground vibration.

Blasting causes generation of shock waves which are similar to seismic waves but the difference is that these waves have lower amplitude, higher dominant frequency so results are not as disastrous as that with earthquakes.

Seismic waves that are generated in ground vibrations are body waves and surface waves.

Body waves are waves that can move within medium whereas Surface waves movement is restricted to surface. Body waves are P waves and S waves and there are number of surface waves like Love waves, Rayleigh wave, stonely waves, channel waves etc.

Body Waves: These are the waves that can travel through the body of earth. These waves originate from the epicentre and move outward. These are of 2 types: P waves and S waves

P Wave or Primary waves or compressional waves: These are waves in form of compression dilation. The movement of particle is in direction of propagation of waves. These are called compressional waves also, and are fastest travelling waves. Due to movement of particles in compression and dilation mode, structure is once in compression and dilation, in this shock waves volume of structure is changed but shape remains the same.

S wave or secondary waves or Shear Waves: S waves are shear waves and particles movement is perpendicular to direction of propagation of wave propagation. On the alignment of movement of particles these waves can be SH (for horizontal alignment) or SV (for vertical alignment). In this shock waves shape of structure is changed but volume remains the same.

Main types of seismic waves

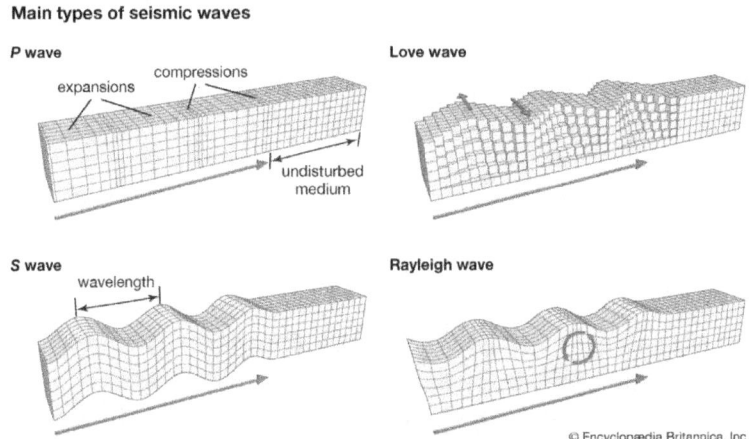

© Encyclopædia Britannica, Inc.

Love Wave
Reyleigh Wave: These waves arise because of constructive interference of P and S waves. Particle movement is elliptical in vertical plane.

Predictor Equation for ground vibration
$PPV = a * (D/Sqrt(Q_{max}))^{-b}$

a and b are site constants and $D/\sqrt{Q_{max}}$ is called scaled distance where D is distance from blast site and Qmax is maximum charge per delay.

If logarithmic plot of same is taken

Log PPV= a- b*SD which gives a linear graph with a being point at SD=0 and b being slope of the curve.

Almost every field engineer engaged in blasting has tried to reach these site constants though this requires consistency and dedication as collecting the number of data and calculating distance from blast hole is done nowadays using GPS apps on mobile. We can calculate distance between 2 points, by marking point A and point B and GPS provides exact vertical and horizontal distance.

What are factors that causes ground vibration

1) Total amount of explosive in blast: Though in the predictor equation there is no mention of total explosive amount used in blasting patch yet there is no denying of the fact that more the amount of explosive, more is the ground vibration produced.

2) Charge per delay: Charge per delay is amount of explosive blasted at one time. Now delay can be provided using NONEL (single charge column firing at a time), or using cord relay or relay detonators (for charging using Detonating Fuse) or in case of Electronic detonator (single charge column at a time). In case of dragline blasting even in case of deck charge, all deck charges will blast together if fired using Detonating Cord, but on the other hand if fired using different delay NONEL, or providing different delay using electronic detonator, explosive charge per delay is significantly

reduced as each decks charge can be given separate delay time for initiation.

If the delay interval of blasting between 2 hole is less than 8ms, the charge per delay would be sum of all those charges that will be blasted in that time zone of 8 ms.

In case of NONEL there is scattering of delay timing. So there are good chances that holes do not blast in the same sequence as they are connected. It may also happen that sometimes holes in back row blast before holes in front row. In such cases there will be more ground vibration.

Improper delay sequencing or scattering will cause issues like back break, generation of boulders (poor fragmentation at some portion).

3) Charging pattern: Type of explosive used also is another factor controlling vibration.

One can also distribute total explosive charge, by positioning deck in explosive column. That would reduce overall quantity as well as reducing maximum charge per delay (if done using NONEL). Again increasing stemming column will decrease amount of charge in hole, but this is not a solution as too much stemming will lead to formation of boulders, so stemming should be optimum. Also if sufficient explosive is not provided, vibration level increases. Blasting should be such that there should be movement of blasted muck. If front row of holes doesn't move then huge energy is reflected back and transmitted back in form of vibration.

Double decking in a single hole is practice in many mines adopted for reduction in ground vibration but if the explosive charge column is not sufficient for breaking the drill pattern, then vibration could increase also.

4) Direction of blast initiation: Direction of blast initiation is good concept to divert vibration from a point. V pattern of firing is very effective in vibration control because in this pattern ground vibration propagation will divert in 2 directions and overall vibration will be reduced, In any firing sequence, direction of propagation of vibration propagation is perpendicularly backwards, to the line of fire.

5) Drill pattern: Vibration depends on the drilling parameters that are used in the blast face. If the charge is reduced such that there is no proper movement of burden (creating new free face for the back rows), then the energy will be dissipated in form of shock waves and increase the ground vibration. Also if the hole is over charged, that will increase the charge per delay and hence vibration.
So drill pattern should be optimum, neither too high nor too low and this should be done by trial and error method.

6) Firing pattern and delay timing: Delay timing and firing pattern is very important for controlling vibration. If delay timing is not correct, and if for any hole effective burden is too high, then vibration will sure increase. Overlapping may cause increase in charge per delay,

thus increasing vibration. Similarly for any blasting, if firing pattern increases effective burden, vibration will spike. Line firing should produce more vibration as compared to diagonal or V pattern or Partial diagonal as in these cases effective burden is reduced. But in staggered holes and if there is good free face line firing should be preferably used.

7) Hardness of strata: These are uncontrollable factors and in general if hardness is more shock waves are more easily transmitted and for longer distances, whereas if there is any discontinuity these shock waves intensity reduces significantly. In softer rocks also absorption of shock energy is high and the shock waves get dissipated faster.

8) Local Geographical characteristics

9) Distance from blast site: The farther is the measuring point lower is the vibration.

Ways to reduce ground vibration:
1. Pre splitting
2. Line drilling
3. Reducing maximum charge per delay by deck.
4. Use of accurate initiation system to reduce scattering
5. Line of fire should be such that vibration direction could be controlled.

Blast Vibration Report and its contents

From the occurrence of one event of ground vibration, the vibrometer records the Peak particle velocities in all three planes i.e. Longitudinal, Transverse and horizontal and it also records air overpressure. The vibrometer consist of Geophones, Linear Mic and recording unit to which these are connected generally referred as vibration monitor. The geophone consists of small plant attached with light spring which oscillates when there is vibration and accordingly graph is plotted for PPV against the time. This graph can be viewed in software. There is another graph plotted against Frequency V/s Time called FFT report.

In event report, one can see the permissible ground vibration limit as per DGMS (India) and plot of PPV vs Frequency. As per DGMS, permissible limit for industrial building, domestic building and historic structure has been set. According to the DGMS permissible limit graph, one can see that for lower frequency, lower value of PPV is permissible, whereas for higher frequency, higher PPV of particles will also work.

Type of Structure	Domina	
	< 8 Hz	8
(A) Buildings/structures not belonging to the owner		
Domestic houses/structures (Kuchha brick and cement)	5	
Industrial Buildings RCC and framed structures)	10	
Objects of historical importance and sensitive structures	2	
(B) Buildings belonging to owner with limited span of life		
Domestic houses/structures (Kuchha brick and cement)	10	
Industrial buildings (RCC & framed structures)	15	

Peak Particle Velocity (PPV) of ground

Seismograms are to be regularly calibrated to ensure that it is measuring accurately ground vibration and air overpressure.

In event Report we encounter ZC frequency.

ZC frequency: Zero crossing frequency is the frequency at which Peak particle velocity is recorded.

The frequency of ground vibration is calculated by measuring the fraction of second that single wave takes and finding its inverse value.

In the plot of PPV, the graph plots the PPV at a given instant. Accordingly Peak Vector Sum (PVS) velocity is also displayed, that takes into consideration of movement of particle in all 3 direction at a given instance. Mic is attached for generating the graph for air overpressure.

In the FFT report, there is plot of amplitude with time and the frequency of vibration at highest amplitude is called dominant frequency.

Fly rocks:

Fly rocks are simply the rocks that are thrown at large distances due to higher amount of energy provided (explosive charged) than required. Flyrocks can be very disastrous, as these can lead to damage of machineries, structure, injury to persons nearby and many other issues. These are result of overcharging, or low drill pattern (depth of hole, or burden or spacing), or holes drilled in loose or blasted area, improper delay timing and hole firing sequence.

Causes of Fly rocks are

> A: Overcharging of Explosive in blast hole
>
> B: Charging of Short Holes
>
> Issue of flyrock is more pronounced in short depth holes. Even when explosive charge is put in controlled amount, then also some flyrocks are yet generated. The biggest issue is when charging coal touch holes as usually these are short depth holes and to add to problems, these holes usually are filled with water. So due to water, stemming is not sufficiently compact, hence flyrocks are increased. In the top benches, where

there is issue of people living in vicinity of mine boundary, we increase the depth of holes thus reducing chances of fly rocks.

C: Charging of holes drilled in loose patch
Another case of fly rocks arise when there is loose strata over which drilling is done.
This is very common practice because full lifting will cause wear to bucket teeth of shovel. Fly rocks can be easily minimized if charging is done while measuring depth of each hole and looking at hole collar, if loose cut explosive accordingly.
D: Top Priming of Explosive: When the explosive charge is initiated by top priming, then the chances of flyrocks increases because of the catering effect of explosive.
E: Short burden or spacing
F: Geological conditions
G: Improper Stemming of holes
H: Improper delay timing and hole firing sequence

At many mines, the issue of fly rock is very critical, as almost all mines are surrounded by human. So to control fly rocks number of techniques are used to prevent fly rocks or stemming ejection.
1: Charging using Electronic detonator: as it gives accurate delay timing which will prevent fly rocks
2: Covering of the hole mouth using sand bags (muffling of the blast patch) and using Steel wire mesh, so that there is no issue of stemming ejection and back fly of rock toward highwall.
3: Controlled charging : Amount of explosive per hole is reduced so as to increase stemming column and to prevent boulder

formation drill pattern is reduced. This helps in prevention of fly rocks.

Air overpressure:

This is caused when there is interaction of energy released in an explosion that comes directly in contact with air. If blasting of explosive is done in open the air nearby is pressurized due to huge generation of gases in blasting. These create pressure which can even be felt as air pressure travelling in form of propagating waves.

Air pressure are described as linear peak overpressure (i.e. increase over atmospheric pressure). Blast monitoring equipment measure the air pressure in unweighted decibels (dB)

4 main causes of air overpressure are

1. Rock pressure pulse: This is generated because of vertical displacement of ground from seismic waves, creating small air pulses. These arrive fastest and have low amplitude and low frequency.

    These arrive almost instantaneously with transverse, longitudinal and vertical ground vibrations.

2. Air Pressure pulse: These are generated because of movement of rock face in vertical or upward direction. This has dominant frequency and is generated due to blasting of each hole. These 2 causes of air overpressure are inevitable and will surely cause air pressure.

3. Gas pressure pulse: This pressure pulse is produced because of leakage of gas pressure generated during blasting into the atmosphere, and connected because of low spacing or due to some discontinuity present in strata.

4. Stemming pressure pulse: This is the case when stemming of blast hole is blown out and blast occurs in contact with air.

If the explosive is blasted in open and is in direct contact with air, there was huge air pressure generated. The air over pressure can be so high that it can cause crack in the machinery glasses and windscreen that are present near the blast patch. Also it was commonly said that number of incidents were found that glass of small at mines cracked because of air pressure.

Air overpressure is measured using Linear Microphone attached with Seismograph. Unit used is in dB (decibel).

Propagation of air blast

As with ground vibration, air blast decays with distance because of geometric spreading. This is the mechanism whereby a finite amount of energy fills an increasing volume of space. There are number of factors that influence air blasts like Topography, weather, wind velocity, temperature, temperature inversions.

Accumulation of explosive in charged sleeping patch on top.

There has been an issue found especially in sleeping patches, that even after stemming is done, then also on next day, some (small amount about 1 -2 kg) explosive comes out of hole and sits on stemming column. This is because of capillary action, in which explosive is pulled up due to formation of gas bubbles, and if stemming is not very compact, small amount of explosive moves up using Nonel, Fuse or Edet. To prevent this case, the rate of gassing must be slower and stemming must be done after giving sufficient time for proper gassing reaction.

There has not been any issue of fragmentation in this case but this phenomenon happens sometimes.

Charging of short holes:

There are some cases when short holes are charged in the blast patch. This issue has severe impact in dragline benches or deep holes benches. In dragline benches, if the short depth blasthole is charged, it may lead to generation of toe, and handling this toe is very difficult.

Even in some patches of Dragline, when full depth holes were not drilled because of loose strata and presence of water which almost instantly collapses the hole collar and charging is very big issue. In that case as soon as drill rod was removed from hole one has to almost charge that hole without the deck keeping some final stemming.

Back Break

This is very common issue and almost every mines has to face this one time or other.

There are number of reasons that can be causative factor for backbreak during a blast.

Few factors that must be considered to prevent backbreak are

1: Delay sequencing and delay scattering

2: Charge column and stemming column

3: Availability of free face

4: Bench stiffness ratio

5: Local geology and presence of fault zone

Crystallization of AN in explosive:

This is not a very common case but sometime AN in emulsion matrix may crystallize out from emulsion and thus the performance of the blast is significantly reduced.
Probable reason for huge crystal formation in emulsion could be because of low quality emulsifier or lower amount of fuel phase in the emulsion. Too much pumping of emulsion in case of transfer, loading unloading can also cause crystal generation in emulsion.

Low Viscosity product

This is very common issue. Whenever the viscosity of explosive is reduced then, we have issues of poor fragmentation in hard benches and in watery holes stemming ejection increases. These issues can be related to uncontrolled gassing (that is very common if viscosity is reduced), but even when final gassing is in limit (nearly 1.1 gm/cc), yet blast performance is not very good.

Lower rate of Gassing

This is another important issue that sometimes become troublesome. This is increasing waiting time in mines and many times this causes blast failure as patches which are charged late, sufficient time is not available. Rate of gassing becomes so slow that even after 30 min of charging there is minimal increase in charge column. This issue usually arises when pH of the product is not in range. For ATC, to catalyze the gassing reaction efficiently, slightly acidic medium is required. For this purpose, pH is maintained nearly at 4.0

Non Initiation of Bulk Explosive due to improper primer (use of Emulsion booster):

Using of low energy primer (especially in Emulsion Booster) can lead to number of blasting issues. The primary reason is if the VOD of primer is lower, then the run for achieving steady state VOD for explosive column increases and significant amount of explosive energy is wasted.

Emulsion Breaking

There is case when the emulsion matrix breaks down separately into oxidizer and fuel phase. There are number of reasons that can lead to such breaking.

1) Lower quality of Raw material containing impurities
2) Poor quality of emulsifier used
3) Rate of blending the oxidizer and fuel phase is too high or too low
4) Presence of rust in blending line as iron content if increases can lead to breaking of emulsion

Alternatives of Rock Blasting

There have been continuous problems due to blasting. There are issues such as ground vibration, flyrocks, noise, generation of fumes and other toxic gases. Beside these direct impacts of blasting, there are other forms of environmental degradation that have been discussed later.

There are some alternatives to the blasting technologies that can be used for removal of ore and rocks, few are mentioned below. The advantage that blasting gives over any other rock breakage technique is low cost and high productivity. Even for a higher rate of production, blasting is the fastest and easiest way.

Blasting can be avoided in soft strata using cutters like in case of coal, limestone use of surface mines, shearers are being deployed, that gives desired fragmented product with good production rate. But in case of hard strata, breakage of rock without blasting, becomes a tiring process. Rippers and expansive chemical agents are required.

1. Eccentric Ripper
2. Expansive chemical agents
3. Highwall mining
4. Use of Surface miners/ coal cutting machines

Expansive Chemical Agents

The primary component of these powders is Lime. The lime when added with water expands on hydration and this property is utilized for generation of cracks in the boulders.

Initially small diameter holes (30-50 mm) are drilled closely in a uniform pattern and then the holes are filled with this slurry. As the slurry expands with time it tends to generate the cracks in the boulders and thus creating big cracks and sometimes breaking the boulders. Development of well defined cracks takes nearly 1-2 days. THis is completely noiseless and without any vibration or other environmental issues. The problem with these powders is this method cannot be fully used in big mines to cope up with the production pressure. These are only used as additional support to the blasting. THese are very useful products currently used in civil construction works, like removal of hard rock that may be encountered during development of land for construction of buildings.

Eccentric Ripper

These are new type of hydraulic breaking hammers. The advantage over the piston hammer is these are more efficient, require less maintenance and generate lower noise.

Highwall mining

This is also a new technology where without full removal of OB, coal is extracted. Initially a trench with depth more than the depth of coal seam to is developed. Then the highwall machine is installed and coal is extracted by cutters. Here one has to leave ribs so as to avoid subsidence of overlying strata. So 100% recovery is not possible.

One great advantage is that after reaching the mine feasibility limit in case of opencast mines, one can get more ore using this technique. Though this method is economical, yet there will be loss of coal as pillars so in long term, this will be no better than rat and pillar mining, where one excavates coal as per his requirements and then leaves the remaining coal to be burnt, this again causing environmental degradation.

Surface Miner/ Coal cutter

These are great machines and are being used as replacement to drilling and blasting, especially in softer strata.

Multiple Choice Questions

1: Stemming Ejection is primarily due to
   a. Loose stemming
   b. Water present in blasthole
   c. Use of Electronic detonator as initiating system
   d. Using detonating fuse as initiating system
Which of the following is wrong:

A: a	B: b
C: c	D: d

Ans: c

2: Stemming ejection is prevented by
   a. Compaction of blastholes using stemming rod
   b. Using bottom initiating of charge column
   c. Increasing the stem column
   d. Increasing Burden in blast design
Which of the above reason is incorrect

A: a	B: b
C: c	D: d

Ans: d

3:Consider the following statement

a. Using the detonating fuse in small diameter hole is not recommended
b. Detonating fuse desensitises small amount of emulsion which is in contact, thus reducing efficiency

Which of the following is correct

A: Both Statement are false	B: a is false but b is true
C: a is true and b is false	D: Both are correct and b explains a

Ans: D

4: Presence of boulders post blasting can be because of?

A: Undercharging of blast holes	B: Variation in strata
C: Presence of preformed boulders	D: Improper delay sequencing

Ans: A, B, C, D

5: Uncontrollable factors that can lead to generation of boulders are

A: Undercharging of blast holes	B: Variation in strata
C: Presence of preformed boulders	D: Improper delay sequencing

Ans: B

6: Handling of boulders are not done using which method?

A: Secondary blasting by drilling holes in the boulder	C: Pop shooting
B: Muffle blasting	D: Crushing rocks using rock breakers
E: Water jet blasting	

Ans: E

7: Overcharging in hole will lead to

A: Chances of flyrocks

B: May Increase in vibration

C: Increase chances of oversized boulders in post blast result

D: Increase in stemming which will lead to air overpressure

Which of the above statement is incorrect

1: A	2: B
3: C	4: D

Ans: 3 and 4

8: Which is the blasting technique used for controlling ground vibration

A: Eccentric Ripper	B: Chemical Blasting
C: Line drilling	D: High wall mining

Ans: C

9: Eccentric ripper is usually not deployed in:

A: Sensitive areas where vibration is to be avoided	B: Very hard massive rocks
C: Soft and medium hard rocks	D: all of these

Ans: B

10: Displacement of particle can be expressed as

A: $a*\sin(wt)$	B: $a*w*\cos(wt)$
C: $a*w*w*\sin(wt)$	D: $a*w*w*\cos(wt)$

Where a is maximum displacement from mean and w is angular velocity

Ans: A

11. Which is most destructive ground vibration among the following

A: PPV=4 mm/s and frequency 5 hz	B: PPV= 4mm/s and frequency of 100 Hz

C: PPV= 2mm/s at Frequncy 5HZ	D: PPV= 2mm/s at frequency of 100 Hz

Ans: A

12. Which of the following waves are body waves

A: Primary waves	B: Secondary Waves
C: Love Waves	D: Stonely Waves

Ans: A and B

13: Which is fastest travelling waves generated in blast vibration

A: P waves	B: S Waves
C: Love Waves	D: Stonely Waves

Ans: A

14: Considering the following properties of waves, which property is not applicable to P waves

A: Movement of particle is in direction of waves propagation	B: Fastest travelling body waves
C: waves propagates in form of compression and dilation	D: These waves have maximum damage impact on structure

Ans: D

15: Consider following statements

1: Body waves are the waves that can travel through inside the earth

2: Love waves and Rayleigh waves are type of body waves

3: P waves and S waves are most destructive waves

4: Body waves travel fastest

Which statement/s are correct

Ans: 1 and 4

16: Predictor equation generally used in mines for prediction of blast

A: PPV= a*(SD)$^{-b}$	B: PPV= a*(SD)*b
C: PPV= a/(SD)*b	D: PPV= a*(SD)

Where SD is Scaled distance and a, b are site constants

Ans: A

17: Use of line drilling

A: Smooth Highwall	B: Vibration control
C: Improved fragmentation	D: Better delay sequencing

Ans: A and B

18: Maximum charge per delay can be reduced using

A: Inserting deck charges in deep hole blasting	B: Using electronic detonator
B: Using proper delay sequencing	D: Using detonating fuse

Ans: A, B, C

19: Probable reason for increasing maximum charge per delay

A: Delay scattering in initiating system	B: Improper blast design
C: Overcharging	D: All the above

Ans: D

20: Pre splitting is beneficial for which conditions

A: Prevention of vibration	B: Stable highwall
C: Reducing flyrocks	D: Smooth highwall post blasting

Ans: A,B, D

21: Which is not component of seismographs

A: Monitor	B: Geophones
C: Linear Microphone	D: Modem

Ans: D

22: Generation of Fly rocks is not related to which of the following factor:

A: Undercharging of blast holes	B: Charging in loose and soft patch
C: Improper delay sequencing	D: Charging of short depth holes

Ans: A

23: Which of the following method are not used for controlling fly rocks

A: Muffling of holes using sand bags and steel wire nets	B: Increasing charge per hole
C: Slightly increasing stemming column	D: Proper hole delay sequencing using accurate delay

Ans: B

24. What can be source of Air Over pressure

A: Rock pressure pulse	B: Air Pressure Pulse
C: Gas pressure pulse	D: Stemming Pressure Pulse

Ans: All the above

Chapter 5: Record Keeping of Blast

Record keeping is necessary evil for any blaster. It is usually considered a boring activity by blaster consuming time and effort. But these are actual assets that a blasting team collects. Technically actual experience a blasting team learns is by noting the details of every day blast on everyday basis. The details that are required for good record keeping are mentioned below

a) Drilling record: Hole depth of all blastholes in a given blast patch along with burden, spacing (drill pattern) and total area of the patch is recorded in drill log sheet.

b) Blasthole Loading sheet: Details of hole depth and explosive charge (bulk and booster/primer charge) per hole are indicated. Beside this there is column indicating water level (if any) and stemming column, initial cup density of Bulk explosive and rate of gassing of explosive (density at interval of 3 min). In case of deck charges, the loading sheet must be representatively showing bottom charge, deck column, top charge and final stemming height along with primer charge in bottom charge column and in top charge column. It must also mention initiating system details.

c) Layout of blastholes (numbered as per loading sheet) indicating surface delay connection and delay sequencing for holes. It must also indicate the cross section of hole roughly indicating charge column in blast hole.

d) Blast report: This is final report generated post blasting.

    a. This report indicates location of blast patch, its RL and distance from set reference point, date and time of blast.

     b.   Total explosive consumption

     c.   Total volume blasted

     d.   Details of drill pattern

     e.   Powder factor achievement

     f.   Maximum charge per delay and distance of seismographs

     g.   Post blast analysis such as flyrock, fragmentation, throw, muckpile profile

e) Vibration Report: These are must for monitoring the vibration level especially if the mines or blasting site is near to anything that is sensitive to blasting (Presence of human, fauna, machineries, structure or industrial buildings, old historic sites etc).

These report and to be maintained and properly filed for the entire mine life. These record not only act as guide for future reference but also very important in case any statutory authority desires to see the records of blast. In case of lawsuits that are filed, these documents are ready references against invalid law suits and claims.

Sample format for each is given below:

Drill log sheet

Loading Sheet and sketch representing stem and deck column and charge column

Patch layout and delay sequencing

Blast report

Vibration report

Sample format for Drill log sheet									
Name of Drilling Partner/Company									
Mines Name					Date				
Location					Shift				
Excavation area/RL					Material				
Patch ID					drill machine				
					Total Drilling Hours				
Drill Machine	Hours	Parameters		upto 6 m	6-9m	9-12m	12m+	Date of Blasting	
		Number of holes							
		Total Depth							
		Number of holes							
		Total Depth							
		Number of holes							
		Total Depth							
		Number of holes							
		Total Depth							
Grant Total		Number of holes							
		Total Depth							
SL No:	Depth	SL No:	Depth	SL No:	Depth	SL No:	Depth	SL No	Depth
1		31		61		91		121	
2		32		62		92		122	
3		33		63		93		123	
4		34		64		94		124	
5		35		65		95		125	
6		36		66		96		126	
7		37		67		97		127	
8		38		68		98		128	
9		39		69		99		129	
10		40		70		100		130	
11		41		71		101		131	
12		42		72		102		132	
13		43		73		103		133	
14		44		74		104		134	
15		45		75		105		135	
16		46		76		106		136	
17		47		77		107		137	
18		48		78		108		138	
19		49		79		109		139	
20		50		80		110		140	
21		51		81		111		141	
22		52		82		112		142	
23		53		83		113		143	
24		54		84		114		144	
25		55		85		115		145	
26		56		86		116		146	
27		57		87		117		147	
28		58		88		118		148	
29		59		89		119		149	
30		60		90		120		150	

Sample Format for Blasthole Loading Sheet

Bulk Explosive Plant							
Date			Mines				
BMD Vehicle			Location				
Hole Diameter			Strata condition				

SL No	depth	Bulk Charge	Booster	Deck	Stem	water
1						
2						
3						
4						
5						
6						
7						
8						
9						
10						
11						
12						
13						
14						
15						
16						
17						
18						
19						
20						
21						
22						
23						
24						
25						
26						
27						
28						
29						
30						
TOTAL		0				

Bulk cup Density	Total Booster
Burden	Total Explosive
Spacing	Length
Average Depth	Width
Volume	Powder factor (m3/kg)

Supplier Signature   Mines representative Signature

Daily Blasting Report

1	Name of Mines	
2	Date of Blast	
3	Place of Blast	
4	Material (OB/Coal)	
5	Name of Magazine	
6	Detail of Blast Parameters	
a)	Number of Holes	
b)	Average depth (in m)	
c)	Average Spacing (in m)	
d)	Average Burden (in m)	
e)	Area of Blast (m2)	
f)	Volume of Blast (in m3)	
g)	Volume of top soil/loose strata (if any)	
h)	Total Explosive consumed (Bulk + Booster) (in kg)	
i)	Maximum Charge per delay (in kg)	
j)	Powder factor (m3/kg)	
k)	PPV (mm/s)	
l)	Frequency of Blast vibration (Hz)	
m)	Location of seismograph	
n)	Distance of Seismograph from Blast Patch (in m)	
o)	Distance of nearest structure from Blast patch (in m)	
p)	Peak sound presssure level	
q)	Fragmentation	
r)	Distance of fly rock projected (in m)	
s)	Throw	
Manufacturer wise details of explosives and accessories consumed		
i)	Accessories company name	
ii)	Total Booster	
iii)	DTH (25/450/11 m) (in nos.)	
iv)	DTH (25/450/9 m) (in nos.)	
v)	DTH (25/450/6 m) (in nos.)	
vi)	TLD 6m (25ms)	
vii)	TLD 6m (42ms)	
viii)	SED (in Nos)	
ix)	Any other Details	

Blasting Overman   Blasting Officer   Mine Manager

Chapter 6: Legal compliances w.r.t. PESO

PESO DEALINGS

Peso stands for Petroleum and explosive safety organization and this government agency came into existence in 1898. This is a government run organization and deals with safety aspects of explosive plant and field operations. It has number of other works also, such as works related with safety related to petroleum and compressed gas. Enforcement of safety aspects related to explosive, petroleum and compressed gas, pressure vessels, gas cylinder, pipelines, LNG, CMG, Ammonium Nitrate, flame proof electrical fittings etc. For handling of explosive division, it is only agency that looks after implementation of Explosive Rules 2008 and Ammonium Nitrate Rules 2012.

Central Headquarters of PESO is at Nagpur, Maharastra.

For explosive plant any change to be done in layout of plant, changing occupier, installation of new facility, addition or detaching of BMD vehicle, changing plant capacity, all must be approved from PESO.

Updating of daily supply figures in PESO

In PESO one has to fill RE -2 form for daily supply of explosive to the respective magazine (for accessories). Each mine gets its accessories for blasting from fixed magazine hence detail of mines are also updated.

Monthly return is filed and uploaded in PESO website and this is to be done by 5th of each month.

PESO deals with every statutory laws and regulations mentioned in Petroleum Act and Explosive Act

Legal compliances w.r.t. Factory Rules

Factory License is governed under Factories Act, 1948. This act covers broad topics on safety and welfare of the workmen involved in the working in an industry. Act empowers inspectors for inspection of the factories and recommends necessary changes in working and takes necessary action which is suited for the workmen involved in the factory.

Legal compliances w.r.t. Pollution (Air and Water) Acts

Pollution control board for hazardous waste disposal and obtaining CTE-CTO.

Pollution control is of utmost importance and every state has government bodies to look after the pollution related issues generated from industries.
State Pollution Control Board implements rules and provision related to following acts:
- Water (Prevention and Control of Pollution) Act, 1974
- The Water (Prevention and Control of Pollution) Cess Act, 1974
- The Air (Prevention and Control of Pollution) Act, 1981
- The Environment (Protection) Act, 1986
- Functions of Central Pollution Control Board :
- Advise the Central Government on matters relating to pollution.
- Coordinate the activities of the State Boards.
- Provide Technical assistance to the State Boards, carry out and sponsor investigations and research relating to control of pollution.

- Plan and organize training of personnel.
- Collect, compile and publish technical and statistical data, prepare manuals and code of conduct.
- To lay down standards.
- To plan a nationwide programme for pollution control.
- Functions of the State Pollution Control Boards:
- To advise the State Government on matter relating to pollution and on siting of industries.
- To plan a programme for pollution control.
- To collect and disseminate information.
- To carry out inspection.
- To lay down effluent and emission standards.
- To issue consent to industries and other activities for compliance of prescribed emission and effluent standards.

Under the provisions of the Water (Prevention & Control of Pollution) Act, 1974 and the Air (Prevention & Control of Pollution) Act, 1981, "any industry, operation or process or an extension and addition thereto, which is likely to discharge sewage or trade effluent into the environment or likely to emit any air pollution into the atmosphere will have to obtain the Consent"

There are two types of the Consent i.e. Consent to Establish (CTE), and Consent to Operate (CTO).

1. Consent to Establish: This consent is required to be obtained before establishing any Industry, Plant, or Process. The Consent to Establish is the primary clearance.
2. Consent to Operate: Once the Industry, Plant, or Process being established according to mandatory pollution control systems, the units are required to obtain consent to operate.

Procedure for obtaining Consent

Under Water and Air Pollution Acts, all industries are classified under Red, Orange, Green and White categories based on their pollution potential and range of pollution index

Documents

While every state has its specific requirement, the following set of documents are usually required across every state for seeking consent under Water Act, 1974 and Air Act, 1981

Consent to Establish
- Site Plan/Location Plan of the industry
- Detailed Project Report which includes the details of raw material, product to be manufactured, the capital cost of the unit (land, building, and plant machinery), water-balance, source of water, and its required quantity
- Land documents such as Registration deed/ Rent deed/Lease deed
- Details of Water Pollution Control/Air Pollution Control instruments
- MOA /partnership Deed

Consent to Operate
- Copy of last consent issued
- Layout plan showing the details of all manufacturing processes
- Latest analysis report of solid waste, effluent, hazardous wastes, and fuel gases
- Copy of balance sheet duly attested by CA or CA certificate
- Detail of land in case the effluent is discharged on land for percolation
- Occupation certificate issued by Town & Country Planning Department, in case of Building & construction projects/area development projects.
- MOA /partnership Deed

Validity and Renewal of Consent

Generally, Consent to Establish is a one-time activity. The State Pollution Control Board issues it for 3 to 5 years. In case the project proponent required an extension of the period, the entrepreneur can apply for an extension basis the requirement.

Consent to operate remains valid for a period of 5, 10, and 15 years according to the red, orange, and green category of the industry respectively.

The industry/project proponent intending for renewal of the Consent to Operate shall apply through OCMMS (Online Consent Management & Monitoring System) before the expiry of the period of previous Consent to Operate permitted by the State Pollution Control Board.

In some states, there is the provision of Auto Renewal in the cases where there isn't any change in their raw material, process, product, increase in overall capital investment cost, and machinery. Production capacity and pollution load of the unit remains the same as declared in the original application for Consent to Operate.

Exemption

The industrial units/projects covered under the white Category are exempted from Consent Management for obtaining CTE and CTO under Water Act, 1974 and Air Act, 1981 and any other units not covered under Red, Orange and Green category.

Penalty

If any Industry / Plant / project operates without obtaining Consent, the entrepreneur shall be liable for imprisonment for a term which may extend to three (3) months or with a fine which may extend to ten thousand rupees (INR 10,000) or with both.

Legal compliances w.r.t. DGMS

DGMS circulars on blasting

Danger Zone: 300m, as per norms but DGMS circular mentions number of cases wherein fly rocks have even reached far more distances than danger zone limit. The actual point that cannot be actually measured is, charging in loose and soft strata, or holes with very low burden. In such cases the only thing that can really be done is proper monitoring of holes during drilling that holes are not drilled in loose strata and the first row of holes from the free face have adequate burden. DGMS should try making rules to prevent drilling in loose, blasted patch, as this is the primary reason for fly rocks, else in true sense these distances are merely of any importance.

Stemming Material for underground mines has been mentioned in DGMS circular for using water or water ampoules, as this will reduce temperature post blasting and significantly reduces chances of fire damp or coal dust explosion, but even in open cast mines benches that if water stemming is used, then blast result is far better than stemming done with clay while displacing water from holes.

Circular for charging of hot holes:
No explosive other than Emulsion and slurry to be used for blasting in hot holes. Blasting should be done using detonating fuse only. If the temperature of hole is above 80deg C, charging of hole will not be done.
There is also a recommendation of DGMS to use emulsion boosters in case of hot holes, instead of PETN based cast

boosters. This is primarily because the emulsion explosives degrade when subjected to higher temperature, instead of detonation.

Danger to ANFO with pyrite ore:
Due to continuous oxidation of pyrite ore, the temperature of holes rises continuously. This has resulted in misfire of ANFO charged holes, hence during charging in pyrite ore mines, with ANFO, 5% urea by weight is added to prevent rise in temperature.

Blasting Time

Shots shall not be fired except during the hours of daylight or until adequate artificial light is provided. All holes charged on any one day shall be fired on the same day as far as practicable.
(ii) as far as practicable, shotfiring shall be carried out either between shifts or during the rest interval or at the end of work for the day.

As per DGMS circular 14 of 2020 dated 24.12.2020, rules for secondary blasting of rocks using explosive have been mentioned.
It mentions that secondary blasting must be avoided and, if at all necessary, small diameter holes of 32 mm must be drilled in the rocks and blasted using cartridge explosives.
Storage of Explosive beyond its shelf life is not permitted and explosive needs to be destroyed.
Instruction for destruction of explosives is issued by office of Chief inspector of Explosive under Cir. 57/1964.
DGMS has issued number of circulars for prevention of pilferage of explosives.

DGMS circular 1 of 2018, dated 13.01.2018, mentions the requirements of safety provisions in Explosive carrier van and explosive charging equipments. Although these provision are mentioned for underground Coal and metalliferous mines:

1. Every explosive carrier should have valid license.
2. Carrier should not be used unless it is fit in condition and complies with explosive rules
3. Electric wiring shall be protected with the conduit. Isolation switch for the battery shall be in accessible position.
4. IT shall not be used for carrying passengers

For explosive charging equipments, the hoses used for pumping of explosives shall be fire resistant and anti static (FRAS).

Also it mentions that if an operator is required to work outside his compartment, working point shall have provision 1. For stopping and starting the engine and

2. for activating the fire suppression system

Environmental Impact

Water Pollution

Emulsion Explosives are manufactured using a number of organic and inorganic raw materials and have harmful impacts to the human body if consumed or inhaled. When the explosive is charged in a blasthole, it comes in contact with the underground water. If the time of contact increases, then leaching of explosives takes place. This will infuse chemicals in the water body and hence poorly affect the person living in nearby areas. To reduce the toxicity of water, the time of sleeping patches must be reduced and load and shoot operation to be preferred.

Air Pollution

During blasting, ideally there is formation of N2, CO2 and water, but in field application, one can see brown fumes, dust generation during blasting. Brown fumes can be indicative of ratio of fuel and oxidizer solution in Emulsion explosive. Yet many other factors like interaction of explosive with in-situ water, higher rate of gassing, presence of impurities in raw material used for manufacturing Emulsion. These fumes must not be inhaled as these may lead to coughing, nausea and choking to the person. Person should only inspect the post blast performance, after the fumes have been removed. Usually the toxic fumes contain CO, CO_2, SO_x, NO_x and other gases. Control of these fumes can be done by maintaining the proper balance but fuel phase and oxidizer solution and rate of gassing should be controlled. Except for these 2 factors, other factors are uncontrollable.

Land Pollution

Waste generated from an emulsion explosive plant must be reused in any form in the blasthole. This is because if the waste is disposed of it will contaminate the land, reduce the fertility of land. Even if a small amount of explosive is not charged inside the blast hole will not detonate and thus by self degradation in some time will contaminate the land.

Emulsions take a long time to degrade also.

Methods of mitigation

Proper utilization of explosives at the field must be ensured.

During manufacturing of emulsion, due care must be taken so that ratio of Fuel oil and Oxidizer solution must be maintained.

Reuse of water and other waste products at manufacturing plants must be ensured and plants must follow the policy of zero discharge of any waste.

The most important statement: The cost of Irresponsibility is huge and not only limited to monetary loss, but loss in all aspects be it materialistic level or mental and spiritual level.